Laser Safety Management

OPTICAL SCIENCE AND ENGINEERING

Founding Editor
Brian J. Thompson
University of Rochester
Rochester, New York

1. Electron and Ion Microscopy and Microanalysis: Principles and Applications, *Lawrence E. Murr*
2. Acousto-Optic Signal Processing: Theory and Implementation, *edited by Norman J. Berg and John N. Lee*
3. Electro-Optic and Acousto-Optic Scanning and Deflection, *Milton Gottlieb, Clive L. M. Ireland, and John Martin Ley*
4. Single-Mode Fiber Optics: Principles and Applications, *Luc B. Jeunhomme*
5. Pulse Code Formats for Fiber Optical Data Communication: Basic Principles and Applications, *David J. Morris*
6. Optical Materials: An Introduction to Selection and Application, *Solomon Musikant*
7. Infrared Methods for Gaseous Measurements: Theory and Practice, *edited by Joda Wormhoudt*
8. Laser Beam Scanning: Opto-Mechanical Devices, Systems, and Data Storage Optics, *edited by Gerald F. Marshall*
9. Opto-Mechanical Systems Design, *Paul R. Yoder, Jr.*
10. Optical Fiber Splices and Connectors: Theory and Methods, *Calvin M. Miller with Stephen C. Mettler and Ian A. White*
11. Laser Spectroscopy and Its Applications, *edited by Leon J. Radziemski, Richard W. Solarz, and Jeffrey A. Paisner*
12. Infrared Optoelectronics: Devices and Applications, *William Nunley and J. Scott Bechtel*
13. Integrated Optical Circuits and Components: Design and Applications, *edited by Lynn D. Hutcheson*
14. Handbook of Molecular Lasers, *edited by Peter K. Cheo*
15. Handbook of Optical Fibers and Cables, *Hiroshi Murata*
16. Acousto-Optics, *Adrian Korpel*
17. Procedures in Applied Optics, *John Strong*
18. Handbook of Solid-State Lasers, *edited by Peter K. Cheo*
19. Optical Computing: Digital and Symbolic, *edited by Raymond Arrathoon*
20. Laser Applications in Physical Chemistry, *edited by D. K. Evans*
21. Laser-Induced Plasmas and Applications, *edited by Leon J. Radziemski and David A. Cremers*

22. Infrared Technology Fundamentals, *Irving J. Spiro and Monroe Schlessinger*
23. Single-Mode Fiber Optics: Principles and Applications, Second Edition, Revised and Expanded, *Luc B. Jeunhomme*
24. Image Analysis Applications, *edited by Rangachar Kasturi and Mohan M. Trivedi*
25. Photoconductivity: Art, Science, and Technology, *N. V. Joshi*
26. Principles of Optical Circuit Engineering, *Mark A. Mentzer*
27. Lens Design, *Milton Laikin*
28. Optical Components, Systems, and Measurement Techniques, *Rajpal S. Sirohi and M. P. Kothiyal*
29. Electron and Ion Microscopy and Microanalysis: Principles and Applications, Second Edition, Revised and Expanded, *Lawrence E. Murr*
30. Handbook of Infrared Optical Materials, *edited by Paul Klocek*
31. Optical Scanning, *edited by Gerald F. Marshall*
32. Polymers for Lightwave and Integrated Optics: Technology and Applications, *edited by Lawrence A. Hornak*
33. Electro-Optical Displays, *edited by Mohammad A. Karim*
34. Mathematical Morphology in Image Processing, *edited by Edward R. Dougherty*
35. Opto-Mechanical Systems Design: Second Edition, Revised and Expanded, *Paul R. Yoder, Jr.*
36. Polarized Light: Fundamentals and Applications, *Edward Collett*
37. Rare Earth Doped Fiber Lasers and Amplifiers, *edited by Michel J. F. Digonnet*
38. Speckle Metrology, *edited by Rajpal S. Sirohi*
39. Organic Photoreceptors for Imaging Systems, *Paul M. Borsenberger and David S. Weiss*
40. Photonic Switching and Interconnects, *edited by Abdellatif Marrakchi*
41. Design and Fabrication of Acousto-Optic Devices, *edited by Akis P. Goutzoulis and Dennis R. Pape*
42. Digital Image Processing Methods, *edited by Edward R. Dougherty*
43. Visual Science and Engineering: Models and Applications, *edited by D. H. Kelly*
44. Handbook of Lens Design, *Daniel Malacara and Zacarias Malacara*
45. Photonic Devices and Systems, *edited by Robert G. Hunsberger*
46. Infrared Technology Fundamentals: Second Edition, Revised and Expanded, *edited by Monroe Schlessinger*
47. Spatial Light Modulator Technology: Materials, Devices, and Applications, *edited by Uzi Efron*
48. Lens Design: Second Edition, Revised and Expanded, *Milton Laikin*
49. Thin Films for Optical Systems, *edited by Francoise R. Flory*
50. Tunable Laser Applications, *edited by F. J. Duarte*

51. Acousto-Optic Signal Processing: Theory and Implementation, Second Edition, *edited by Norman J. Berg and John M. Pellegrino*
52. Handbook of Nonlinear Optics, *Richard L. Sutherland*
53. Handbook of Optical Fibers and Cables: Second Edition, *Hiroshi Murata*
54. Optical Storage and Retrieval: Memory, Neural Networks, and Fractals, *edited by Francis T. S. Yu and Suganda Jutamulia*
55. Devices for Optoelectronics, *Wallace B. Leigh*
56. Practical Design and Production of Optical Thin Films, *Ronald R. Willey*
57. Acousto-Optics: Second Edition, *Adrian Korpel*
58. Diffraction Gratings and Applications, *Erwin G. Loewen and Evgeny Popov*
59. Organic Photoreceptors for Xerography, *Paul M. Borsenberger and David S. Weiss*
60. Characterization Techniques and Tabulations for Organic Nonlinear Optical Materials, *edited by Mark G. Kuzyk and Carl W. Dirk*
61. Interferogram Analysis for Optical Testing, *Daniel Malacara, Manuel Servin, and Zacarias Malacara*
62. Computational Modeling of Vision: The Role of Combination, *William R. Uttal, Ramakrishna Kakarala, Spiram Dayanand, Thomas Shepherd, Jagadeesh Kalki, Charles F. Lunskis, Jr., and Ning Liu*
63. Microoptics Technology: Fabrication and Applications of Lens Arrays and Devices, *Nicholas Borrelli*
64. Visual Information Representation, Communication, and Image Processing, *edited by Chang Wen Chen and Ya-Qin Zhang*
65. Optical Methods of Measurement, *Rajpal S. Sirohi and F. S. Chau*
66. Integrated Optical Circuits and Components: Design and Applications, *edited by Edmond J. Murphy*
67. Adaptive Optics Engineering Handbook, *edited by Robert K. Tyson*
68. Entropy and Information Optics, *Francis T. S. Yu*
69. Computational Methods for Electromagnetic and Optical Systems, *John M. Jarem and Partha P. Banerjee*
70. Laser Beam Shaping, *Fred M. Dickey and Scott C. Holswade*
71. Rare-Earth-Doped Fiber Lasers and Amplifiers: Second Edition, Revised and Expanded, *edited by Michel J. F. Digonnet*
72. Lens Design: Third Edition, Revised and Expanded, *Milton Laikin*
73. Handbook of Optical Engineering, *edited by Daniel Malacara and Brian J. Thompson*

74. Handbook of Imaging Materials: Second Edition, Revised and Expanded, *edited by Arthur S. Diamond and David S. Weiss*
75. Handbook of Image Quality: Characterization and Prediction, *Brian W. Keelan*
76. Fiber Optic Sensors, *edited by Francis T. S. Yu and Shizhuo Yin*
77. Optical Switching/Networking and Computing for Multimedia Systems, *edited by Mohsen Guizani and Abdella Battou*
78. Image Recognition and Classification: Algorithms, Systems, and Applications, *edited by Bahram Javidi*
79. Practical Design and Production of Optical Thin Films: Second Edition, Revised and Expanded, *Ronald R. Willey*
80. Ultrafast Lasers: Technology and Applications, *edited by Martin E. Fermann, Almantas Galvanauskas, and Gregg Sucha*
81. Light Propagation in Periodic Media: Differential Theory and Design, *Michel Nevière and Evgeny Popov*
82. Handbook of Nonlinear Optics, Second Edition, Revised and Expanded, *Richard L. Sutherland*
83. Polarized Light: Second Edition, Revised and Expanded, *Dennis Goldstein*
84. Optical Remote Sensing: Science and Technology, *Walter Egan*
85. Handbook of Optical Design: Second Edition, *Daniel Malacara and Zacarias Malacara*
86. Nonlinear Optics: Theory, Numerical Modeling, and Applications, *Partha P. Banerjee*
87. Semiconductor and Metal Nanocrystals: Synthesis and Electronic and Optical Properties, edited by *Victor I. Klimov*
88. High-Performance Backbone Network Technology, *edited by Naoaki Yamanaka*
89. Semiconductor Laser Fundamentals, *Toshiaki Suhara*
90. Handbook of Optical and Laser Scanning, *edited by Gerald F. Marshall*
91. Organic Light-Emitting Diodes: Principles, Characteristics, and Processes, *Jan Kalinowski*
92. Micro-Optomechatronics, *Hiroshi Hosaka, Yoshitada Katagiri, Terunao Hirota, and Kiyoshi Itao*
93. Microoptics Technology: Second Edition, *Nicholas F. Borrelli*
94. Organic Electroluminescence, *edited by Zakya Kafafi*
95. Engineering Thin Films and Nanostructures with Ion Beams, *Emile Knystautas*
96. Interferogram Analysis for Optical Testing, Second Edition, *Daniel Malacara, Manuel Sercin, and Zacarias Malacara*
97. Laser Remote Sensing, *edited by Takashi Fujii and Tetsuo Fukuchi*
98. Passive Micro-Optical Alignment Methods, *edited by Robert A. Boudreau and Sharon M. Boudreau*

99. Organic Photovoltaics: Mechanism, Materials, and Devices, *edited by Sam-Shajing Sun and Niyazi Serdar Saracftci*
100. Handbook of Optical Interconnects, *edited by Shigeru Kawai*
101. GMPLS Technologies: Broadband Backbone Networks and Systems, *Naoaki Yamanaka, Kohei Shiomoto, and Eiji Oki*
102. Laser Beam Shaping Applications, *edited by Fred M. Dickey, Scott C. Holswade and David L. Shealy*
103. Electromagnetic Theory and Applications for Photonic Crystals, *Kiyotoshi Yasumoto*
104. Physics of Optoelectronics, *Michael A. Parker*
105. Opto-Mechanical Systems Design: Third Edition, *Paul R. Yoder, Jr.*
106. Color Desktop Printer Technology, *edited by Mitchell Rosen and Noboru Ohta*
107. Laser Safety Management, *Ken Barat*
108. Optics in Magnetic Multilayers and Nanostructures, *Stefan Visnovsky*

Laser Safety Management

Ken Barat

CRC Taylor & Francis
Taylor & Francis Group
Boca Raton London New York

A CRC title, part of the Taylor & Francis imprint, a member of the Taylor & Francis Group, the academic division of T&F Informa plc.

Published in 2006 by
CRC Press
Taylor & Francis Group
6000 Broken Sound Parkway NW, Suite 300
Boca Raton, FL 33487-2742

Printed in the United States of America on acid-free paper
10 9 8 7 6 5 4 3 2 1

International Standard Book Number-10: 0-8247-2307-4 (Hardcover)
International Standard Book Number-13: 978-0-8247-2307-1 (Hardcover)
Library of Congress Card Number 2005052919

Library of Congress Cataloging-in-Publication Data

Barat, Kenneth.
Laser safety management / Kenneth Barat.
p. cm. -- (Occupational safety and health ; 107)
Includes bibliographical references and index.
ISBN 0-8247-2307-4
1. Lasers--Safety measures. I. Title. II. Occupational safety and health (Taylor & Francis) ; 107.

TA1677.B36 2005
621.36'60289--dc22 2005052919

informa
Taylor & Francis Group
is the Academic Division of Informa plc.

Visit the Taylor & Francis Web site at
http://www.taylorandfrancis.com

and the CRC Press Web site at
http://www.crcpress.com

Preface

This book came out of a late-night meeting between Ken Barat and John O'Hagan at the Laser Institute of America's International Laser Safety Conference in 2003. Both of us have been involved with training courses in laser safety for many years, but on separate continents. Since that time, we have shared experiences and views on laser safety training, and this book tries to bring together those experiences. While John had to pull out of the final product, much of his work, words, and insight are contained in this book. I will always think of this as our book.

Mention laser safety to many people and their immediate reaction is that this is something that is going to cost time, money, and effort. It is something that gets in the way of research and profit. Most appreciate that lasers can present safety issues; we have the popular perception of the "ray gun" to thank for that. Very few see the positive benefits of implementing a practical laser-safety management program. We make a bold claim at this stage: A well-thought-through laser-safety management program will save you time, money, and effort. It will improve the quality of your research program. It will improve your reputation as an organization, and, ultimately, it can improve the bottom line. While the approach and many examples demonstrated in this book are geared toward laser use in research and development or academic settings, it is easily translated to any laser use environment.

How can we make these claims? We make them based on experience with applying the techniques in this book across many types of organizations in a number of countries. There is some pain to be suffered in setting things up, but like many things in life, the effort is worth it.

There are many excellent books on the theory and practical operation of lasers, so we have not tried to repeat these. This is a practical guide to laser-safety management techniques. Whether you are a chief executive officer, general safety officer, laser safety officer, postgraduate researcher, or manufacturing supervisor, this book should provide you with the tools to implement a practical laser-safety management program.

The laser was first successfully demonstrated in 1960 and has become a ubiquitous part of our life, in both the workplace and the home. Compared with a lot of other technology, the number of deaths and reported injuries involving laser technology is relatively small. Does this suggest that there is no problem and managing laser safety is a waste of time? We suggest not. Laser safety has generally been a success story because the technology arrived at a time when caution would have been expected. Most of those people at risk of injury in the early days were those who developed the technology. By the 1970s we had manufacturing safety standards for laser products, which coincided with the first public experiences of lasers in supermarkets for scanning barcodes.

Many of us use laser products in the office and home without having to consider laser safety. The laser printer and compact disc player are just tools; the manufacturer has (we hope) ensured that we can safely use these devices.

An explosion of incidents involving laser pointers highlighted the concern over laser products being in the hands of the public. Many of these incidents hit the headlines in the media and possibly fuèled further incidents. The perception of those who had the lasers, which were suddenly both very available and very cheap, was: How could such a small device possibly be harmful? The perception of those who were targeted was that some sort of death ray was blinding them. The situation was not helped by conflicting messages from laser safety and security "experts." What was not taken into account was the availability of an amazing piece of technology — way beyond the wildest dreams of those early researchers — that could teach a whole new generation the wonders of laser beams. We will try to address the perceived conflict of easy access with ensuring safe use as one application of the laser-safety management program.

An organization buying a laser product to carry out a specific activity may only need to consider most of the laser-safety management issues up to the implementation stage. For others, especially those doing research, laser-safety management may be a daily issue.

Let us consider the impact of having a laser safety incident in the workplace. It does not really matter whether the workplace is a manufacturing plant or a university research laboratory. For our hypothetical incident we could have a pair of researchers working in a laboratory aligning a laser beam through a specific path on an optical bench. We will assume that the laser beam is relatively low power and is green. One researcher is adjusting a mirror while the other is watching a small screen on which he is expecting the laser beam to appear. It does not, so he turns around and at that moment his colleague moves the mirror too far and targets him momentarily in the eye with the beam. What happens next?

Has the recipient of the beam suffered an eye injury or is he just dazzled? Do either of the two researchers understand the risk posed by the laser beam? Do they know what to do now? They could call for help. There could be bitter feelings because eyewear was not worn, why alignment aids were not used, or more fundamentally, why the operator was not more careful? Perhaps punctuated by more colorful language. Emotions are likely to take over for a while. Certainly there will be an element of guilt and concern on the part of the researcher doing the adjustment. The recipient of the beam may be on his way to the hospital. Let us leave them for the moment and think about the wider impact.

The research has stopped. It may have been part of a very important, well-funded and urgent project. The safety office and managers have been notified and an investigation begins, followed almost immediately by a lot of blame shifting. This takes time that could have been spent doing other things. Perhaps the regulators need to be informed. They stop all laser work at the facility until they have carried out their investigation. The problem now affects the work of many other researchers. Within a relatively short time the local newspaper has heard that something has happened and reporters start arriving. The insurance company representative arrives and informs management that they will not meet any claim because the researchers were either not properly trained or were inadequately supervised.

The regulators decide to prosecute the establishment. This now becomes bigger news and generates a new wave of media interest. One knock-on effect of this is that the funder of the research becomes uncomfortable with the media attention and threatens to withdraw the funding.

Although the "casualty" does have an injury on the retina, it is far from the critical central vision region and does not affect the quality of his life (too much). He decides to sue for compensation.

The management now starts to take an interest in the use of this "dangerous" laser equipment in the establishment. There are two options: get rid of all of the lasers or try to implement a proactive laser-safety management program.

Summarizing the "cost": someone was injured; his fellow researcher and many staff members suffered stress; a lot of time was consumed at every level of the organization; there was a financial penalty for the fine and increased insurance premiums; and research across the organization was delayed. One additional major cost was to the reputation of the organization. This can be the most difficult to recover from.

This is perhaps a dramatic story, but it is based on a real incident. The conclusion of the real incident was the implementation of a very effective laser-safety management program. It has become part of the research mindset, and the quality of research has improved.

When you look at who is generally responsible for ensuring an adequate level of laser safety, the list is quite long. We can focus on the organization and work our way down from the top, from managers with specific responsibility for general safety and perhaps specifically for laser safety to the users of the laser equipment. However, the net is much wider than this. We can branch out to the suppliers of the equipment and through them to the manufacturers and designers. There may also be consultants who advise all parties along the way.

Our objective with this book is to provide practical tools for all who are responsible for laser safety. You will read many of the suggestions and perhaps think, "Well that is just common sense." You will be right, but sometimes even the obvious has to be spelled out. We will refer you to other resources where relevant. Some of the details of laser-safety management are continuously evolving. Please contact me to share your experiences — and suggestions that do not work. This is how we have learned over the years. The methodologies covered will work in any workplace.

As already alluded to, this book would not have happened without a lot of people making mistakes and a few doing well. We have learned from them all and extend our thanks to them. The participants in our laser-safety training courses have taught us a lot of what we know, as have those to whom we have given advice. Our peers in laser safety have humbled us with their commitment and dedication to the application of laser technology in a safe way.

Our hope was to make this book as useful as possible regardless of where in the world one uses lasers.

Author

Ken Barat is currently the laser safety officer (LSO) for the National Ignition Facility (NIF) Directorate at Lawrence Livermore National Laboratory (LLNL). He oversees the laser safety for the NIF project (the world's largest laser project) and all program laboratories (60+), performing audits, training, and developing courses and other means to ensure that all laser users are safe. His goal is to find that compromise between safety and getting the work done: "safety through cooperation."

He is a fellow of the Laser Institute of America, second winner of the Jim Rockwell Award (2005) for Laser Safety Leadership and the Tim Renner User Service Award (2002) from the Advance Light Source. He is the former LSO for Lawrence Berkeley National Lab (LBNL). As former group leader of the non-ionizing group of the Arizona Radiation Regulatory Agency, he successfully led the effort for statewide laser regulations. He is also one of the founders of the Bay Area Laser Safety Officer Society (BALSO), the nation's oldest active LSO networking group. He was among the first LSOs certified by the Board of Laser Safety.

He has given presentations on laser safety at national and international meetings and acted as a consultant and trainer for a wide varity of companies: Motorola University, Chrysler, Spectra Physics, N-light, University of California, Imperial College, U.K., RIOS, Hewlett Packard, Beckin Dickinson, Coherent Inc, Laser Institute of America, Health Physics Society, Rockwell Laser Industries, and NIST.

He is a member of the United Kingdom's Laser Safety Forum and helped launch and instructed both the Hong Kong University of Science and Technology's first laser safety seminars and the Hong Kong Medical Society's LSO training seminar. He has numerous publications on laser safety. He has been the seminar director for laser safety programs for LBNL, LLNL, and Stanford University. He is an active member on several an ANSI committees. He has a degree in chemistry and is a product of New York City (Brooklyn) public education.

Contents

Dedication

Just like at an award ceremony, I have many special people to thank for making this book happen: my family (Leah, Emily, and my wife, Pat) for moral support; my parents Lil and Mac Barat for shaping me; John O'Hagan for his professionalism; the encouragement of the Bay Area Laser Safety Officer Society; a special thanks to Jim Rockwell and Matt Kowoskoy for seeing something in me.

Disclaimer

This book is not a product of the Department of Energy, University of California, Lawrence Livermore National Laboratory, National Ignition Facility Directorate.

1 Overview

1.1 TRADITIONAL LASER SAFETY

The traditional way that laser safety has been addressed for decades is laid out in the American National Institute Standard for the Safe Use of Laser Z136.1. This can be found in the Hazard Evaluation and Classification section of the standard.

Several aspects of the application of a laser or laser system influence the total hazard evaluation and thereby influence the application of control measures:

1. The laser or laser system's capability of injuring personnel or interfering with task performance
2. The environment in which the laser is issued
3. The personnel who may use or be exposed to laser radiation

Preceding this hazard evaluation protocol is the appointment of a laser safety officer or advisor, who will be referred to in this text as the LSO. It is the LSO's responsibility to see that laser safety is adequately addressed at a facility or institution.

1.2 LIFE CYCLE

A more comprehensive evaluation approach is to expand this evaluation from three components to five and then add two life cycle phases: design and disposal. The five components are:

1. The laser or laser system's capability of injuring personnel
2. The beam path of the laser system
3. The interaction of the laser beam with its intended target
4. The environment in which the laser is used
5. The personnel who may use or be exposed to the laser radiation

1.2.1 The Laser or Laser System's Capability of Injuring Personnel

Here are the factors of the laser source itself:

1. What type of laser? Here one is interested in details such as whether it uses a pulsed or continuous wave (CW) and the nature of the

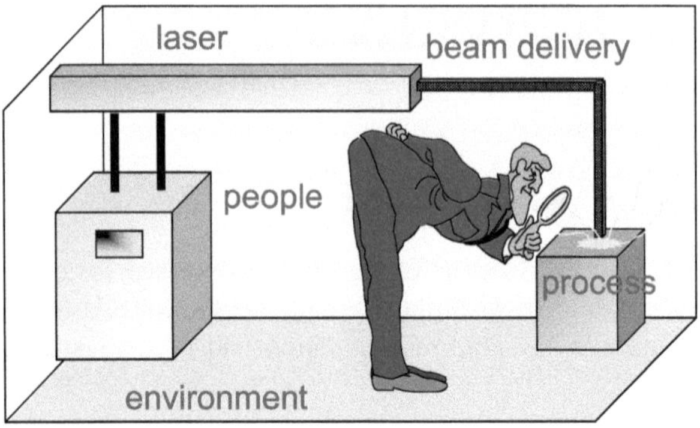

FIGURE 1.1 Representation of complete hazard evaluation approach.

wavelengths being generated, that is, ultraviolet, visible, near infrared, mid- or far infrared? Each of these will have some effect on the nature and level of control measures and hazard the laser posses to an individual.

2. What is the output of the laser? Are we talking about milliwatts, nanojoules, megawatts, or joules of output? These answers and the wavelength will have a dramatic effect on possible laser protective eyewear requirements.
3. What is the classification of the laser? In the research and development (R&D) environment almost all lasers are class 3R(3A), 3B, or 4.

1.2.2 The Beam Path of the Laser System

You need to consider what happens to the laser beam once it leaves the laser source. In a very similar way you consider your commute from leaving the security of your garage to your work destination. Are you one of the lucky ones who has a minor commute of several minutes, or do you have a long arduous commute of highways, tunnels, and bridges?

In laser terms the beam path could be open, contained in fiber optics, or enclosed. In addition, just like the driver going down a steep grade, the beam could be amplified or go through nonlinear optics and therefore produce a change of lanes in our driving example, but for photons it is a change of wavelength. This could occur several times along with possible chirp stretching or compression. Any of these steps or a combination of them will affect the safety requirements one might apply to a system.

1.2.3 The Interaction of the Laser Beam with its Intended Target

Once the laser radiation reaches its destination, just like our driver reaching work, many options lie ahead, from that great day at work to violent meetings. A percentage of the beam may be reflected off a target, or beam interaction may generate gases as a result of products requiring ventilation. An intense pulse laser beam may generate ionizing radiation in the form of neutrons; gamma or x-rays even cause activation of products, hence generating additional ionizing radiation. Maybe the end of the beam path is delivered through a robotic arm, which introduces new concerns for evaluation.

1.2.4 The Environment in which the Laser is Used

Now we have to consider factors from the workplace and how they contribute to our hazard evaluation. Do they make our job easier or harder? Places such as a clean room may do both. While adding to access control and thereby helping keep unauthorized persons out, cleanness requirements may make it harder to implement other controls. Other common laser use environments are the operating room, manufacturing floor, fabrication area, and our chief interest, the research laboratory.

1.2.5 The Personnel Who May Use or be Exposed to Laser Radiation

When we think of these people, authorized laser users, ancillary staff, visitors, and in some case consumers come to mind. As each of these groups is evaluated, items such as training requirements, personnel protective equipment, and even ergonomic factors require consideration.

1.3 ADDITIONAL LIFE CYCLE ELEMENTS

1.3.1 Design

The designing of safety into a laser product is a clear legal responsibility of the laser manufacturer, as called out in national product safety codes and regulations. Items such as protective housings, interlocks, labels, and electrical safety controls are all rather standard and expected by the purchaser of such products. The reader should check the Web site of the Center of Radiological Devices and Radiological Health for a listing of requirements and guidance documents. The design of a research setup is more a by-product of environmental or experimental need to reduce air turbulence or keep out unwanted light pollution than part of an overall safety plan. In some settings the laser beam path may travel across several optical tables and even across walkways or through a wall. Time given to providing a safe work environment will pay dividends to the user and those visiting the laser use area. Designers also need to think about what goes into the laser and related equipment and how to dispose of it.

1.3.2 Disposal

Few laser safety professionals think about disposal of laser equipment until the issue is brought before them. Unlike radioactive material or radiation-generating products, there is little control over who buys laser products or how they are disposed of. This may be changing. Laser products can contain hazardous materials whose disposal can require special care. This is highlighted by the international effort to rid electronic and other products of hazardous materials, that is, the "greening of products." Regulations that require this are Waste from Electrical and Electronic Equipment (WEEE) and Restriction of Hazardous Substances (RoHS). Both of these European norms and others push the concept of corporate social responsibility, which fits very well into the design and disposal phases of our approach.

Some of the common ways to dispose of laser equipment rather than sending it to landfills are:

1. Donating the equipment to educational institutions
2. Contacting resale firms
3. Looking for a home for the equipment within one's own organization
4. Auctioning the equipment
5. Returning it to the manufacturer (Some manufacturers do have limited return for disposal programs).

1.4 CLASSIFICATION

Laser hazard classification gives the user or laser safety officer an initial sense of the hazard the laser system or product presents to the user and others in the area. Rather than using colors to alert one to the fire hazard level in laser safety, a numerical code is used. The higher the number, the greater the hazard potential. Potential is the key word, for any laser system can be made safe. The hazard levels range from class 1, no hazard, to class 4, maximum potential hazard (Table 1.1). ANSI, the International Electrotechnical Commission (IEC), and the Center for Devices and Radiological Health (CDRH) each had slightly different classification systems until 2005, when they all adopted a uniform approach.

A key component of laser safety is the hazard classification scheme (Figure 1.2), which is an indication of the laser's capability of injuring personnel. All laser or laser systems are classified according to their accessible radiation during operation, which in a research setting can be different from the classification of the laser source. Thus, a class 3B laser beam can be amplified on an optical table to class 4. Likewise, a class 3B or 4 laser beam can be attenuated to a lower classification as part of an optical set up. Laser products sold in the United States are usually classified in accordance with the Federal Laser Product Performance Standard, which falls under the Food and Drug Administration (FDA), CDRH. Those sold in Europe are classified to meet IEC 60825-1. Be aware that under CDRH Laser Notice 50, laser products sold in the United States can also be labeled with a certification as meeting IEC 60825-1. If the laser has been modified subsequent to classification by the manufacturer, its falls upon the LSO to classify the new laser system or product.

Lasers are classified according to their potential to cause biological damage. The pertinent parameters are laser output energy or power, radiation wavelengths,

TABLE 1.1
Laser Hazard Classification

Class	Basis for Classification
Class 1: Safe Visible and nonvisible	Lasers that are safe under reasonably foreseeable conditions of operation; generally a product that contains a higher-class laser system but access to the beam is controlled by engineering means.
Class 2: Low power Visible only	For CW lasers, protection of the eyes is normally provided by the natural aversion response, including the blink reflex, which takes approximately 0.25 sec. (These lasers are not *intrinsically* safe.) AEL = 1 mW for a CW laser.
Class 1M: Safe without viewing aids 302.5 to 4000 nm	Safe under reasonably foreseeable conditions of operation. Beams are either highly divergent or collimated but with a large diameter. May be hazardous if user employs optics within the beam.
Class 2M: Safe without viewing aids Visible only	Protection of the eyes is normally provided by the natural aversion response, including the blink reflex, which takes approximately 0.25 sec. Beams are either highly divergent or collimated but with a large diameter. May be hazardous if user employs optics within the beam.
Class 3R: Low and medium power 302.5 nm to 1 mm	Risk of injury is greater than for the lower classes but not as high as for class 3B. Up to 5 times the AEL for class 1 or class 2.
Class 3B: Medium and high power Visible and nonvisible	Direct intrabeam viewing of these devices is always hazardous. Viewing diffuse reflections is normally safe provided the eye is no closer than 13 cm from the diffusing surface and the exposure duration is less than 10 sec. AEL = 500 mW for a CW laser
Class 4: High power Visible and nonvisible	Direct intrabeam viewing is hazardous. Specular and diffuse reflections are hazardous. Eye, skin and fire hazard. Treat class 4 lasers with caution.

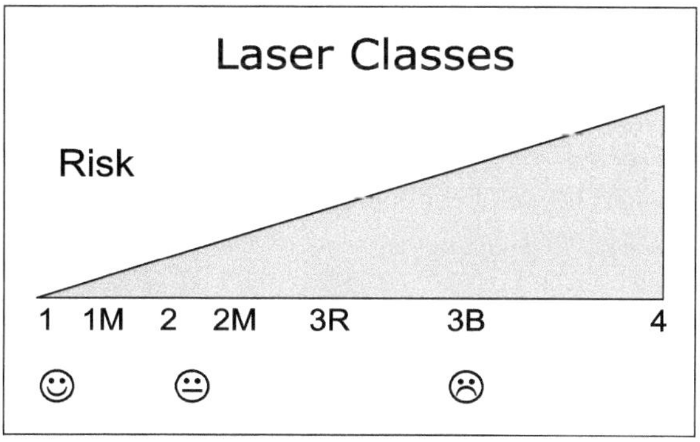

FIGURE 1.2 Risk vs. hazard classification.

exposure duration, and cross-sectional area of the laser beam at the point of interest. In addition to these general parameters, lasers are classified in accordance with the *accessible emission limit* (AEL), which is the maximum accessible level of laser radiation permitted within a particular laser class.

In laser safety standards and regulations, laser hazard classifications are used to signify the level of hazard inherent in a laser system and the extent of safety controls required. These range from class 1 lasers (which are inherently safe for direct beam viewing under most conditions) to class 4 lasers (which require the most strict controls). The laser classifications are described as follows:

1.4.1 Time Base for Classification

Different time bases are used for the different classes and wavelength ranges as follows:

1. 0.25 s for class 2, class 2M, and class 3R in the wavelength range from 400 to 700 nm
2. 100 s for laser radiation of all wavelengths above 100 nm, excepted for cases listed in items 1 and 3
3. 30,000 s for laser radiation of all wavelengths less than or equal to 400 nm and for laser radiation greater than 400 nm where intentional long-term viewing is inherent in the design or function of the laser product

1.4.1.1 Class 1

Any laser or laser system that cannot cause eye or skin injury during normal operation qualifies as class 1. They cannot emit laser radiation at known hazard levels (typically CWs of 0.4 μW at visible wavelengths). Users of class 1 laser products are generally exempt from radiation hazard controls during operation and maintenance (but not necessarily during service). The maximum exposure duration is assumed to be no more than 30,000 sec, except that for infrared systems not intended to be viewed (>0.7 μm), 100 sec must be used. The exemption strictly applies to emitted laser radiation hazards and not to other potential hazards.

1.4.1.2 Class 1M

They are safe under reasonably foreseeable conditions of operation but may be hazardous if observed using viewing optics. Examples of this class are Light Emitting Diodes (LED) and fiber communication systems.

Two hazardous conditions apply:

1. If the beam is diverging and someone uses (for example) a lens within 100 mm of the aperture to collimate or otherwise concentrate the beam into the eye
2. If the beam is of a large diameter and collimated and someone uses a lens to increase the proportion of the beam that can enter the eye

1.4.1.3 Class 2

These are low-power visible lasers that emit above class 1 levels but emit a radiant power not above 1 mW. The concept is that the human aversion reaction to bright light will protect a person. Aversion or blink response is less than 0.25 sec. A CW HeNe laser above class 1 but not exceeding 1 mW radiant power is an example of a class 2 laser. The laser safety professional should be aware that counter to common belief, studies have indicated the 0.25 sec blink reflex may be present in less than 25% of the general population.

1.4.1.4 Class 2M

Class 2M lasers (safe if not using viewing aids) are restricted to the wavelength range of 400 to 700 nm. Protection of the eyes is normally provided by the aversion response. The difference between a class 2 and a class 2M laser is that the total power in the beam of a class 2M laser can be much higher. However, the beam will either be highly divergent or collimated and have a large diameter so that the proportion of the beam that can normally enter the eye is small. Class 2M lasers are generally not safe if viewing optics are used.

Two hazardous conditions apply:

1. If the beam is diverging and someone uses (for example) a lens within 100 mm of the aperture to collimate or otherwise concentrate the beam into the eye
2. If the beam is of a large diameter and collimated and someone uses a lens to increase the proportion of the beam that can enter the eye

1.4.1.5 Class 3R

This class consists of lasers and laser systems that have an accessible output between 1 and 5 times the class 1 AEL for wavelengths shorter than 0.4 μm or longer than 0.7 μm, or less than 5 times the class 2 AEL for wavelengths between 0.4 and 0.7 μm, meaning 1 to 5 mW. The "R" stands for reduced requirements. As with class 3A below, which is only hazardous for intrabeam viewing, class 3R can be considered safe for momentary viewing except by optics.

1.4.1.6 Class 3A

This is the predecessor of class 3R: intermediate-power lasers (CW: 1 to 5 mW). Such lasers are only hazardous for intrabeam viewing and can be considered safe for momentary viewing except by optics. Some limited controls are usually recommended.

NOTE: There are different labeling requirements for class 3a lasers with a beam irradiance that does not exceed 2.5 mW/cm^2 (caution logotype) and those where the beam irradiance does exceed 2.5 mW/cm^2 (danger logotype).

1.4.1.7 Class 3B

These are moderate-power lasers, invisible wavelengths: cannot generate radiant energy greater than 125 mJ, CW: 5–500 mW, pulsed: in less than 0.25 sec., visible wavelengths: CW 5–500mW, pulsed cannot produce radiant energy greater than 30 mJ per pulse. In general, class 3B lasers will not be a fire hazard nor are not generally capable of producing hazardous diffuse reflections except for conditions of intentional staring done at distances close to the diffuser. Specific controls are recommended. The application of these controls should be graded since a direct exposure from a 10 mW laser does not present the hazard a direct exposure would from a 400 mW laser. The diffuse reflection from both might not present any hazard at a distance greater than 0.1 meters.

1.4.1.8 Class 4

High-power lasers (CW: 500 mW, pulsed capable of generating over 125 mJ in less than 0.25 sec) are hazardous to view under any conditions (directly or diffusely scattered) and are a potential fire hazard and a skin hazard. Significant controls are required of class 4 laser facilities. A graded approach may be justified.

Class 1 Product: ***A laser system or product that contains a completely enclosed laser or laser system of a high classification.***

The international system is defined in the IEC laser safety standard IEC 60825-1. This has been adopted by a number of countries within their national standards. The United States and ANSI z136 has recently adopted this system.

1.5 RESPONSIBILITIES

While it is common to think of the responsibility for laser safety as the sole responsibility of the LSO, the responsibility really falls among many levels.

1.5.1 Management

The employer will provide employment and a safe and healthful place of employment for his employees. This statement comes from the regulation for the California Occupational Safety and Health Administration. Similar wording and expectations can be found in many regulatory sources. Part of management's responsibility for providing a safe work place, from a laser safety perspective, is the appointment of an LSO. This appointment does not have to be, and generally is not, a full-time position. Once the appointment is made, the employer through its management chain takes on certain responsibilities. Quoting the ANSI Z136.1 2000 version Section 5.3.2.1, "the management shall provide for training to the LSO on the potential hazards, control measures, applicable standards, and any other pertinent information pertaining to laser safety and applicable standards or provide the LSO adequate consultative services. The training shall be commensurate to at least the highest-class laser under the jurisdiction of the LSO; Training also includes consideration for non-beam hazards."

1.5.2 Laser Safety Officer or Advisor

The role of the LSO is critical to ensuring laser safety, particularly in the R&D setting. Without an appointed LSO, an organization cannot say it addresses laser safety. The LSO is commonly defined as an individual with the authority and responsibility to monitor and enforce the control of laser hazards. In addition, the LSO must be able to knowledgeably evaluate and control laser hazards. The LSO either performs the stated tasks or ensures that the tasks are performed.

Some of the LSO's responsibilities are listed below:

1. Classification
2. Hazard evaluation
3. Control measures
4. Procedure approval
5. Protective equipment
6. Signs and labels
7. Facility and equipment
8. Safety feature audits
9. Training
10. Medical surveillance
11. Record keeping
12. Accident investigation
13. Review of laser system operations

1.5.3 Research Group Leader or Supervisor

This individual, depending on the organization, is known by several names: principal investigator, responsibility individual, senior researcher, group leader, or supervisor. It is common for these individuals to not see themselves as supervisors or want the responsibilities that go along with the title. As in so many things, the attitude is set from the top down. If management or the group leader only present lip service to laser safety, but no time or resources, soon everyone will sense how unimportant laser safety is to management and their actions will reflect this. Several kinds of laser accidents can be traced back to management's lack of belief in or support of laser safety.

1.5.4 Employees

No matter how detailed or creative a laser safety program may be, if the employees do not see a reason for laser safety, the laser safety system will fail. The employees must understand that they play the most critical role in laser safety, particularly in the R&D setting. Some items they must feel free to do are:

1. Bring real-life issues to the attention of group leadership or the LSO.
2. Have the flexibility to get corrections or modifications made to work procedures.
3. Be aware that complaining to oneself or to peers generally does not produce any positive changes. There must be a mechanism to raise concerns to a higher level.
4. Report accidents or problems without fear of their careers being negatively impacted.

1.5.5 Field Service Staff and Vendors

These people need to realize the vital role they play in laser safety. They are looked upon by many to be the system experts and therefore the source of laser safety controls and guidance. A casual word or outright disregard for laser safety can have disastrous effects.

1.6 PROBABILISTIC RISK ASSESSMENT

For some technologies, such as nuclear power generation and chemical plants, people want to know they are going to be safe. The problem is that ultimate safety cannot be achieved, so generally techniques are used to determine if certain events or actions are likely, the results from these events or actions are determined, and numbers are assigned. How do these numbers evolve? Do we need to go to these lengths when using our laser product? Of course, it depends on what can go wrong and the consequences.

We could use the same formal techniques developed for other technologies, and there are many — for example, probabilistic risk assessment, fault-tree analysis, HAZOP, and so on. These often make use of data from real incidents and accidents. The media like to quote the probability of us being killed per mile driven, riding in a train, and flying in an airplane. Of course, behind these figures are, tragically, real cases of people who have been killed using these modes of transport. Does information exist for people who have been injured or killed using your particular laser product? It may, but probably not. What if your laser product is unique and part of a research project? How can you evaluate it? In general, it turns out that we can go a long way toward assessing safety issues without even switching the laser on.

As an illustration, we can carry out a thought experiment — so favored by Albert Einstein, but we will be a bit simpler. Banana skins have provided a source of amusement in slapstick comedy for at least a century. Can we find out the probability of someone being killed by slipping on a banana skin? We can define our probability here as the ratio of the number of people killed to the number of people who could have slipped on the banana skin or the ratio of the number of people killed to the number of people who actually slipped, which shows us the first difficulty. We need to be careful to understand what numbers mean.

We could do the actual experiment. We could take the population of a small town, say of 1,000 people, and ask them one by one to walk along somewhere where we have left a banana skin. We would have to make sure that the volunteers could not communicate with each other and could not see what was happening; otherwise they could bias our experiment (by learning from other people's experience — for once something we do not want to happen). At the end of the experiment, we could count up the number of people who:

1. Stepped over or to the side of the banana skin
2. Slipped on the banana skin but suffered no fall or injury
3. Slipped on the banana skin, fell, and either:
   a. Were not injured
   b. Received minor injuries
   c. Received major injuries
   d. Received fatal injuries

There are a number of problems with trying to do this experiment, not least of which is the potential litigation you could face. It would also be extremely difficult to do the experiment with a single community without your volunteers finding out about the fate of earlier participants. However, within national databases of accidents and emergencies you could almost certainly find some real data.

So, how do we get a gut feeling? You probably already have it. Without doing the experiment, you could probably think through — from your experience of everyday life — the probability of being killed by slipping on a banana skin. Your estimate is probably as good as any official statistics. You could also try a comparison of estimates with colleagues or friends over a coffee or beer.

Is there a correct answer to how many people will be killed? Not until you do the actual experiment. Even then, the answer is only correct for that experiment. For 1,000 people, the estimated answer is 0 or 1. If you thought 5, you would not be wrong. Of course it is possible that all 1000 people will be killed. There may be reasons why the likelihood of slipping and death is higher. The participants could be blindfolded and, through careful design of the route they are instructed to take, the likelihood of slipping could be practically certain. The banana skin could also be positioned at the edge of a 100-m drop off a cliff, increasing the probability of death.

We can use exactly the same thought experiment technique to assess our laser application. However, we also need to think about risk assessment. We all need to be very good at real, practical risk assessment in our everyday lives. A natural selection process exists for those who are not good at risk assessment. As an example, we can take the process we go through when crossing a road, generally subconsciously.

As you approach the edge of the roadway, you start the process. You may use all of the senses available to you, but you will primarily use your eyes and ears. Essentially, you start looking at the traffic on the road — but what are you looking for? It might be the volume of traffic, its speed, and the type of traffic.

You may take into account other factors to do with the environment, including how far along the road you can see and whether the traffic is all moving in the same direction. Let us concentrate on the traffic. What does this consist of? There could be cars, vans, trucks of various sizes, and perhaps bicycles.

This initial process is hazard spotting. Hazards are defined as the things that can cause harm. We are interested in the harm to people, essentially us at this stage. We may take into account who may be harmed by the hazard and how they could be harmed. If it is us crossing the road then obviously we are at risk of something happening to us. However, we may have other people with us. A group of fit young adults will primarily look after themselves but will (generally) still keep a lookout for others in the group. However, an adult with young children will do the risk assessment for the whole group. We will define risk as the result of combining the probability that a hazard will do us harm and the outcome from that interaction. Interacting with a high-speed heavy truck can have drastic consequences for us; we could be spread over a large part of the roadway. However, we may expect to survive an interaction with a bicycle, and cars are somewhere in between.

Our perception of the risk will depend on a number of factors, not least on our past experience. If we have crossed this particular road many times before with no near-misses, we may become complacent. If something has changed — for instance, the traffic is moving much faster than usual — we may exercise more caution.

We also need to consider our tolerance of the risk. Let us assume that you are crossing the road to catch a bus. You have been offered a new job where your salary is to be doubled, but you have to get there by a certain time. The bus has just arrived at the stop on the other side of the road. The traffic is very busy. What do you do? Do you wait for a reasonable opportunity to cross and risk missing the bus, or do you make a dash between the traffic? Under these circumstances you may well expose yourself to a much greater risk because you perceive that a benefit could result. Crossing under the same traffic conditions at other times would be an intolerable risk to you.

Going back to standing at the side of the road, there are a number of things we could do. We could decide not to cross the road. That way, we have removed the risk because we will not be exposed to the hazard. We could wait for a suitable gap in the traffic and cross. We may or may not know how long this will take. We could look for something to help us cross the road. Perhaps the first is a pedestrian crossing. This is only effective if the drivers actually stop. We could call such a control measure an administrative control measure because it is procedural. You are told when learning to drive that you should stop at such crossings, but nothing in the vehicle control system forces you to do so. If the crossing has lights, then this may be a bit better. The driver is now being guided by a control system (rather than necessarily observing you), but still there is nothing to physically prevent the driver from running you over. This is an improvement on the administrative control measure, but it is by no means perfect.

How else could we get to the other side of the road? There may be a pedestrian bridge. This may mean walking farther and the additional effort of climbing steps or walking up a ramp. There is obviously the potential risk of a driver hitting any supports for the bridge, or even a vehicle approaching that is too tall to fit under the bridge. Generally, the probability of these events is small, but they are possible.

Another approach is to use an underpass or tunnel. The probability of a vehicle dropping down through the roof is very unlikely, but is there other potential problems. In some neighborhoods and at some times of the day (or more accurately, night) you may consider that the risk of death by mugging is much greater than the risk of being run over by a truck.

Both the bridge and the tunnel are examples of engineering control measures. They both separate us from the hazard so that the risk of harm from the hazards is very small. However, the control measure may introduce other risks, as is the case with the tunnel.

We build up our experience of crossing various roads throughout life and we call on that experience all of the time. If you see an incident involving someone crossing a road at a particular location, that event may come to mind whenever you cross at that location.

Let us assume we are standing at the side of a relatively quiet road. A cyclist is approaching and he is wearing a bright green helmet. He is not particularly close and you judge that since the traffic is otherwise clear, you will now cross the road. Just as you step onto the road surface the cyclist speeds up and appears to try to run you down. However, you make it to the other side of the road and the cyclist speeds off into the distance. The following day you come to the same place and are ready to cross the road. You see a cyclist approaching with a bright green helmet. What goes through your mind? You may stand back and let him past or you may take some other action; it depends on your personality. It may not even be the same cyclist as yesterday. What you have done is learned from the experience and modified your assessment.

When you cross the road for real, this whole assessment process takes place very quickly and without much conscious thought. We can summarize the process as follows:

1. Spot the hazards.
2. Decide who may be harmed by the hazards and how they could be harmed.
3. Identify existing control measures and assess the residual risk.
4. Record the findings from the assessment.
5. Periodically review the assessment.

These are the so-called five steps to risk assessment, which are commonly applied to laser safety in the United Kingdom and are slowly gaining a foothold in the United States. The above example shows that risk assessment can be based on common sense. All of this feeds back into the five steps of the comprehensive hazard evaluation that we started the chapter with.

2 Biological Effects: Why We Care About Laser Exposure

Do not stare at laser with remaining eye!

2.1 INTRODUCTION

The chief concern over laser use has always been the possibility of eye injury. While skin presents a larger target, it is the possibility of injury to the eyes that drives laser safety, funding, controls, and application. The effect of laser radiation varies with the wavelength and the part of the eye it interacts with. In addition, biological effects from direct exposure and diffuse reflection exposure differ. This chapter explains the anatomy of the eye and skin and issues associated with biological effects.

2.2 THE EYE

The major danger of laser radiation is from beams entering the eye. The eye is the organ most sensitive to light. A laser beam with low divergence entering the eye can be focused down to an area 10 to 20 μm in diameter.

A 40-mW laser is capable of producing enough energy (when focused) to instantly burn through paper. The energy density (measure of energy per unit of area) of the laser beam increases as the spot size decreases. This means the energy of a laser beam can be intensified up to 100,000 times by the focusing action of the eye for visible and near-infrared (NIR) wavelengths. If the irradiance entering the eye is 1 mW/cm^2, the irradiance at the retina will be 100 W/cm^2. Thus, even a low-power laser in the milliwatt range can cause a burn if focused directly onto the retina.

Light from an object (such as a tree) enters the eye first through the clear cornea and then through the pupil, the circular aperture (opening) in the iris. Next, the light is converged by the lens to a nodal point immediately behind the lens; at that point, the image becomes inverted. The light progresses through the gelatinous vitreous humor and, ideally, back to a clear focus on the retina, the central area of which is the macula. In the retina, light impulses are changed into electrical signals and then sent along the optic nerve and back to the occipital (posterior) lobe of the brain, which interprets these electrical signals as visual images.

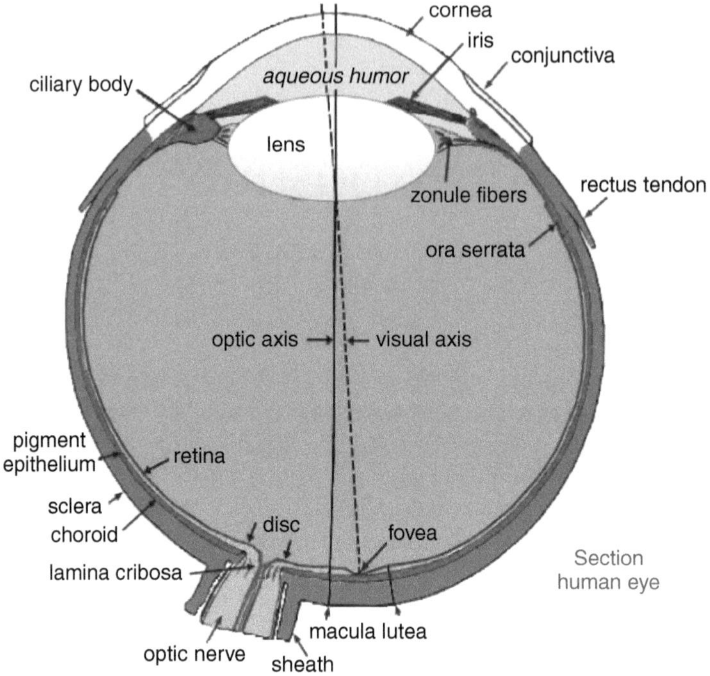

FIGURE 2.1 (**See color insert following page 164.**) Human eye components.

Damage to the eye is dependent upon the wavelength of the beam. In order to understand the possible health effects, it is important to understand the functions of the major parts of the human eye, as shown in Figure 2.1.

2.2.1 Parts of the Human Eye

2.2.1.1 The Cornea

The cornea is the transparent layer of tissue covering the eye. Damage to the outer cornea may be uncomfortable (like a gritty feeling) or painful but will usually heal quickly. Damage to deeper layers of the cornea may cause permanent injury.

2.2.1.2 The Lens

The lens focuses light to form images onto the retina. Over time, the lens becomes less pliable, making it more difficult to focus on near objects. With age, the lens also becomes cloudy and eventually opacifies. This is known as a cataract. Every lens develops cataracts eventually.

2.2.1.3 The Retina

The part of the eye that provides the most acute vision is the fovea centralis (also called the macula lutea). This is a relatively small area of the retina (3 to 4%) that provides the most detailed and acute vision as well as color perception. This explains why eyes move when you read; the image has to be focused on the fovea for detailed perception. The rest of the retina perceives light and movement. If a laser burn occurs on the fovea, most fine (reading and working) vision may be lost. If a laser burn occurs in the peripheral vision, it may produce little or no effect on vision.

Four kinds of light-sensitive receptors are found in the retina: rods and three types of cones. Each cone is tuned to absorb light from a portion of the spectrum of visible light, long-wavelength light (red), middle-wavelength light (green), and short-wavelength light (blue). Each type of receptor has its own special pigment for absorbing light. Each consists of a transmembrane protein called opsin, which is coupled to the prosthetic group retinal. Retinal is a derivative of vitamin A and is used by all four types of receptors.

2.2.2 Blink and Aversion Response

Fortunately the eye has a self-defense mechanism: the blink and aversion response. Aversion response is the closing of the eyelid or movement of the head to avoid exposure to bright light. The aversion response is commonly assumed to occur within 0.25 sec and is only applicable to visible laser wavelengths. This response may defend the eye from damage where low-power lasers are involved but cannot help where high-power lasers are involved. With high-power lasers, the damage can occur in less than a quarter of a second.

A study by Reidenbach et al.* states that even for the larger spot size on the retina, the frequency of the blink reflex has been shown to be less than 35%, and the same is true for the maximum of the pupil size, that is, for low ambient light conditions. In this study 503 volunteers were irradiated in the lab and 690 in 4 different field trials with laser radiation. Out of these only 15.5% and 18.26%, respectively, showed a blink reflex. The respective numbers as a function of wavelength are 15.7% (670 nm), 17.2% (635 nm), and 22.4% (532 nm) for long-, middle-, and short-wavelength light. An increase from 4.2% to 28.1% in blink reflex frequency was achieved when the ambient illuminance was decreased from 1700 lx to lx 1 using an LED as a large extended stimulating optical source instead of a collimated laser beam.

A further dependency was found concerning the irradiated area on the retina. Increasing the retinal spot from 6.4 to 9.4 mm^2 to 33.7 to 46.8 mm^2 resulted in an increase of the blink reflex percentage from 20 to 33.3%.

---

* Reidenback, H.D., Hofmann, J., Dollinger, K., and Seckler, M. A Critical Consideration of the Blink Reflex as a Means for Laser Safety Regulations. International Laser Safety Conference, 2005.

2.2.3 Where Laser Radiation Goes

2.2.3.1 Ultraviolet-B and Ultraviolet-C (100 to 315 nm)

The surface of the cornea absorbs all ultraviolet (UV) ranges of these wavelengths, which produce a photokeratitis (welders flash) from a photochemical process, which causes a denaturation of proteins in the cornea. This is a temporary condition because the corneal tissues regenerate very quickly, less than 24 hours.

2.2.3.2 Ultraviolet-A (315 to 400 nm)

The cornea, lens, and aqueous humor let in UV radiation of these wavelengths, and the principal absorber is the lens. Photochemical processes denature proteins in the lens, resulting in the formation of cataracts.

2.2.3.3 Visible Light and Infrared-A (400 to 1400 nm)

The cornea, lens, and vitreous fluid are transparent to wavelengths. Damage to the retinal tissue occurs by absorption of light, and its conversion to heat occurs by the melanin granules in the pigmented epithelium or by photochemical action to the photoreceptor. The focusing effects of the cornea and lens increase the irradiance on the retina by up to 100,000 times. For visible light of 400 to 700 nm, the aversion reflex, which takes 0.25 sec, may reduce exposure by causing the subject to turn away from a bright light source. However, this will not occur if the intensity of the laser is great enough to produce damage in less than 0.25 sec or when light of 700 to 1400 nm (NIR) is used, as the human eye is insensitive to these wavelengths.

2.2.3.4 Infrared-B and Infrared-C (1400 to $1.0 \times 10 + 6$ nm)

Corneal tissue will absorb light with a wavelength longer than 1400 nm. Damage to the cornea results from the absorption of energy by tears and tissue fluids, causing a temperature increase and subsequent denaturation of protein in the corneal surface (Figure 2.2).

2.3 SIGNS OF EYE EXPOSURE

Symptoms of a laser burn in the eye include a headache shortly after exposure, excessive watering of the eyes, and sudden appearance of floaters. *Floaters* are swirling distortions that occur randomly in normal vision, usually after a blink or when eyes have been closed for a couple of seconds. Floaters are caused by dead cells that detach from the retina and choroid and float in the vitreous humor. Ophthalmologists often dismiss minor laser injuries as floaters because of it is difficult to detect minor retinal injuries. Minor corneal burns cause a gritty feeling, like sand in the eye.

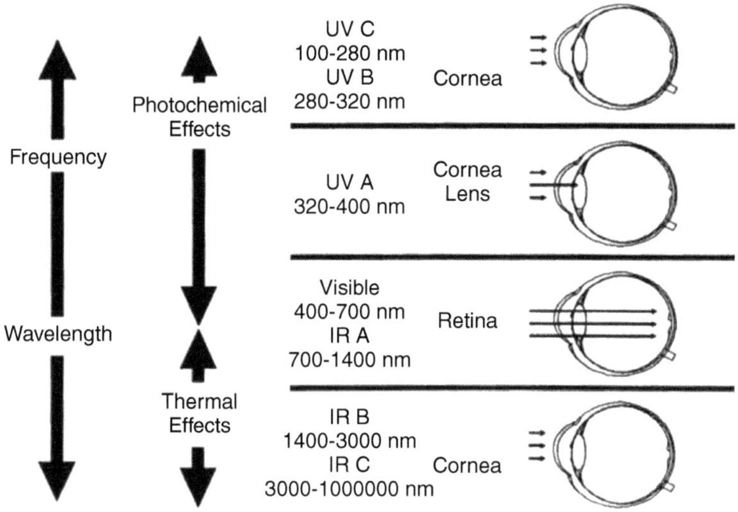

FIGURE 2.2 Transmission and damage mechanism by wavelength.

Exposure to a visible laser beam can be detected by a bright color flash of the emitted wavelength and an after-image of its complementary color (e.g., a green 532-nm laser light would produce a green flash followed by a red after-image). When the retina is affected, there may be difficulty in detecting blue or green colors secondary to cone damage, and pigmentation of the retina may be detected.

Exposure to the Q-switched Nd:YAG laser beam (1064 nm) is especially hazardous and may initially go undetected because the beam is invisible and the retina lacks pain sensory nerves. Photoacoustic retinal damage may be associated with an audible pop at the time of exposure. Visual disorientation caused by retinal damage may not be apparent until considerable thermal damage has occurred.

Exposure to the invisible carbon dioxide laser beam (10,600 nm) can be detected by a burning pain at the site of exposure on the cornea or sclera. Table 2.1 summarizes the hazards caused by various light wavelengths.

2.4 DAMAGE MECHANISMS

2.4.1 Electromechanical and Acoustic Damage

This type of damage requires beams of extremely high-power density (10^9 to 10^{12} W/cm^2) in extremely short pulses (nanoseconds) to delivery fluences of about 100 J/cm^2 and very high electric fields (106 to 107 V/cm), comparable to the average atomic or intermolecular electric field. Such a pulse induces dielectric breakdown in tissue, resulting in a microplasma or ionized volume with a large number of electrons. A localized mechanical rupture of tissue is caused by the shock wave associated with the plasma expansion. Laser pulses of less than 10 μsec can induce a shock wave in the retinal tissue that causes tissue

TABLE 2.1
Hazards Caused by Various Light Wavelengths

Wavelength Range	Effect on Eye	Effect on Skin
Ultraviolet C 200–280 nm	Photokeratitis	Erthema (sunburn) Skin cancer Accelerated skin aging
Ultraviolet B 280–315 nm	Photokeratitis	Increased pigmentation
Ultraviolet A 315–400 nm	Photochemical cataract	Pigment darkening Skin burn
Visible 400–700 nm	Photochemical Thermal retinal injury	Pigment darkening Skin burn
Near-infrared 700–1,400 nm	Cataract and retinal burn	Skin burn
Mid-infrared 1,400–3,000 nm	Corneal burn Aqueous flare, cataract	Skin burn
Far-infrared 3,000–100,000 nm	Corneal burn	Skin burn

rupture. This damage is permanent, as with a retinal burn. Acoustic damage is more destructive to the retina than a thermal burn. Acoustic damage usually affects a greater area of the retina, and the threshold energy for this effect is substantially lower. The ANSI MPE values are reduced for short laser pulses to protect against this effect.

2.4.2 Photoablation

Photoablation is the photodissociation, or direct breaking, of intramolecular bonds in biopolymers, caused by absorption of incident photons and subsequent release of biological material. Molecules of collagen, for example, may dissociate by absorption of single photons in the 5- to 7-eV energy range. Excimer lasers at several UV wavelengths (ArF, 193 nm/6.4 eV; KrF, 248 nm/5 eV; XeCl, 308 nm/4 eV) with nanosecond pulses focused on tissue at power densities of about 10^8 W/cm^2 can produce this photoablative effect. UV radiation is extremely strongly absorbed by biomolecules, and thus absorption depths are small, of the order of a few micrometers.

2.4.3 Thermal Damage

Thermal damage is caused by the conversion of laser energy into heat. With the laser's ability to focus on points a few micrometers or millimeters in diameter, high power densities can be spatially confined to heat target tissues. Depth of penetration into the tissue varies with wavelength of the incident radiation, determining the amount of tissue removal and bleeding control.

The photothermal process occurs first with the absorption of photon energy, producing a vibrational excited state in molecules, and then in elastic scattering with neighboring molecules, increasing their kinetic energy and creating a rise in temperature. Under normal conditions the kinetic energy per molecule (kT) is about 0.025 eV. Heating effects are largely controlled by molecular target absorption such as free water, hemoproteins, melanin, and other macromolecules such as nucleic acids.

2.4.4 Photochemical Damage

Light below 400 nm does not focus on the retina. The light can be laser output, UV from the pump light, or blue light from a target interaction. The effect is cumulative over a period of days. The ANSI standard is designed to account only for exposure to laser light. If UV light from a pump light or blue light from a target interaction is emitted, additional precautions must be taken.

2.5 LASER RADIATION EFFECTS ON SKIN

Laser radiation injury to the skin is normally considered less serious than injury to the eye, since functional loss of the eye is more debilitating than damage to the skin, although the injury thresholds for both skin and eyes are comparable (except in the retinal hazard region (400 to 1400 nm)). In the far-IR and far-UV regions of the spectrum, where optical radiation is not focused on the retina, skin injury thresholds are about the same as corneal injury thresholds. Obviously, the possibility of skin exposure is greater than that of eye exposure because of the skin's greater surface area.

The layers of the skin that are of concern in a discussion of laser hazards are the epidermis and the dermis. The epidermis layer lies beneath the stratum corneum and is the outermost living layer of the skin. The dermis mostly consists of connective tissue and lies beneath the epidermis. Figure 2.3 shows the three layers of the skin.

2.5.1 Epidermis

The epidermis is the outer layer of skin. The thickness of the epidermis varies in different types of skin. It is the thinnest on the eyelids, at .05 mm, and the thickest on the palms and soles of the feet, at 1.5 mm. The epidermis is made up of five layers. From bottom to top the layers are stratum basale, stratum spinosum, stratum granulosum, stratum licidum, and stratum corneum. The bottom layer, the stratum basale, has cells that are shaped like columns. In this layer the cells divide and push already formed cells into higher layers. As the cells move into the higher layers, they flatten and eventually die. The top layer of the epidermis, the stratum corneum, is made of dead, flat skin cells that shed about every 2 weeks. There are three types of specialized cells in the epidermis. The melanocyte produces pigment (melanin), the Langerhans cell is the frontline

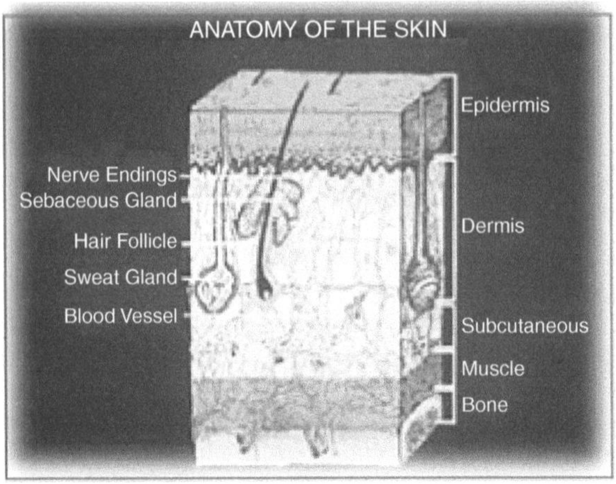

FIGURE 2.3 Labeled cross section of human skin.

defense of the immune system in the skin, and the Merkel cell's function is not clearly known.

2.5.2 Dermis

The dermis also varies in thickness depending on the location of the skin. It is 0.3 mm on the eyelid and 3.0 mm on the back. The dermis is composed of three types of tissue that are present throughout, not in layers. The types of tissue are collagen, elastic tissue, and reticular fibers.

The two layers of the dermis are the papillary and reticular layers. The upper, papillary, layer contains a thin arrangement of collagen fibers. The lower, reticular, layer is thicker and made of thick collagen fibers that are arranged parallel to the surface of the skin.

The dermis contains many specialized cells and structures. The hair follicles are situated here with the erector pili muscle that attaches to each follicle. Sebaceous (oil) glands and apocrine (scent) glands are associated with the follicle. This layer also contains eccrine (sweat) glands, but they are not associated with hair follicles. Blood vessels and nerves course through this layer. The nerves transmit sensations of pain, itch, and temperature. There are also specialized nerve cells called Meissner's and Vater-Pacini corpuscles that transmit the sensations of touch and pressure.

2.5.3 Subcutaneous Tissue

The subcutaneous tissue is a layer of fat and connective tissue that houses larger blood vessels and nerves. This layer is important in the regulation of temperature of the skin and the body. The size of this layer varies throughout the body and from person to person.

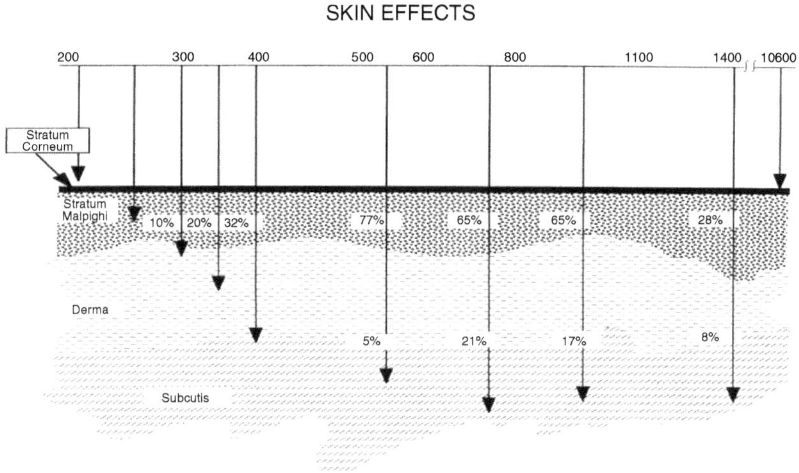

FIGURE 2.4 Skin transmissions by wavelength. Safety with Lasers & Other Optical Sources, Sliney & Walbarsht.

Figure 2.4 shows the depth of penetration into the skin for different wavelengths of laser radiation. There is quite a variation in depth of penetration over the range of wavelengths, with the maximum occurring around 700 to 1200 nm. Injury thresholds resulting from exposure of less than 10 seconds to the skin from far-IR and far-UV radiation are superficial and may involve changes to the outer dead layer of the skin. A temporary skin injury may be painful if sufficiently severe, but it will eventually heal, often without any sign of injury. Burns to larger areas of the skin are more serious, as they may lead to serious loss of body fluids. Hazardous exposure of large areas of the skin is unlikely to be encountered in normal laser work.

A sensation of warmth resulting from the absorption of laser energy normally provides adequate warning to prevent thermal injury to the skin from almost all lasers except for some high-power far-IR lasers. Any irradiance of 0.1 W/cm^2 produces a sensation of warmth at diameters larger than 1 cm. One-tenth of this level can be readily sensed if a large portion of the body is exposed. Long-term exposure to UV lasers causes long-term delayed effects such as accelerated skin aging and skin cancer.

To the skin, UV-A (0.315 to 0.400 μm) can cause hyperpigmentation and erythema. UV-B and UV-C, often collectively referred to as "actinic UV," can cause erythema and blistering, as they are absorbed in the epidermis. UV-B is a component of sunlight that is thought to have carcinogenic effects on the skin. Exposure in the UV-B range is most injurious to skin. In addition to thermal injury caused by ultraviolet energy, there is the possibility of radiation carcinogenesis from UV-B (0.280 to 0.315 mm) either directly on DNA or from effects on potential carcinogenic intracellular viruses.

Exposure in the shorter UV-C (0.200 to 0.280 μm) and the longer UV-A ranges seems less harmful to human skin. The shorter wavelengths are absorbed in the outer dead layers of the epidermis (stratum corneum), and the longer wavelengths have an initial pigment-darkening effect followed by erythema if there is exposure to excessive levels. IR-A wavelengths of light are absorbed by the dermis and can cause deep heating of skin tissue.

2.6 TISSUE OPTICS

An introduction to tissue optics will serve the reader well. Optical properties of skin reflect the structure and chemical composition of the skin. When a beam of light reaches the skin surface, part of it is specularly reflected by the surface, while the rest is refracted and transmitted into the skin.

The light transmitted into the skin is scattered and absorbed by the skin. After multiple scattering, some of the transmitted light will reemerge through the air–stratum interface into the air. This reemergence is called *diffuse reflection.* The amount of diffuse reflection is determined by the scattering and absorption properties of the skin tissue. The stronger the absorption, the less the diffuse reflection; the stronger the scattering, the larger the diffuse reflection. Following absorption of a photon by the skin, the electrically excited absorbing molecule may rapidly return to a more stable energy state by re-emission of a photon with lower energy, that is, fluorescence emission.

Figures 2.5 through 2.11 show examples of biological damage. There are several major contributors to the absorption spectrum. In the UV, absorption increases with shorter wavelength because of protein, DNA, and other molecules. In the infrared, the absorption increases with longer wavelengths because of tissue water content. In the red to NIR absorption is minimal.

Whole blood is a strong absorber in the red to NIR regime, but because the volume fraction of blood is a few percent in tissues, the average absorption coefficient that affects light transport is moderate. However, when photons strike a blood vessel they encounter the full strong absorption of whole blood. Hence, local absorption properties govern light–tissue interactions, and average absorption properties govern light transport.

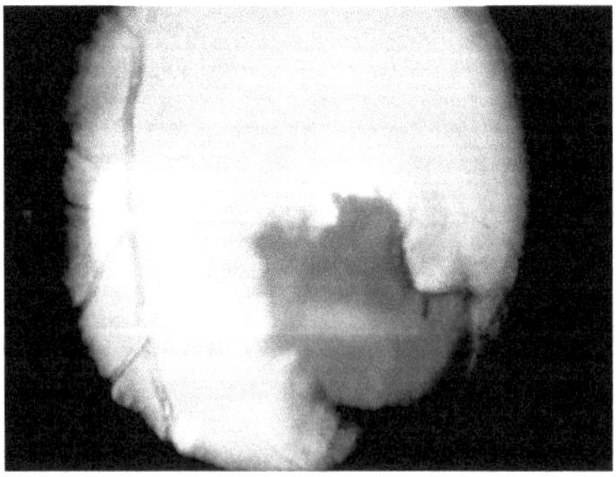

FIGURE 2.5 **(See color insert following page 164.)** Blood in vitreous.

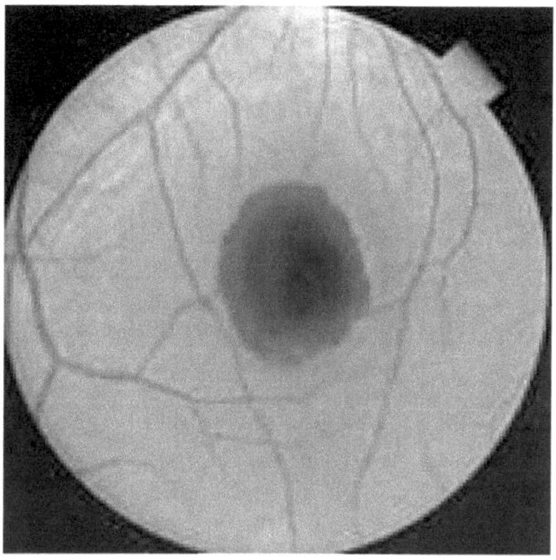

FIGURE 2.6 **(See color insert following page 164.)** Pooled blood.

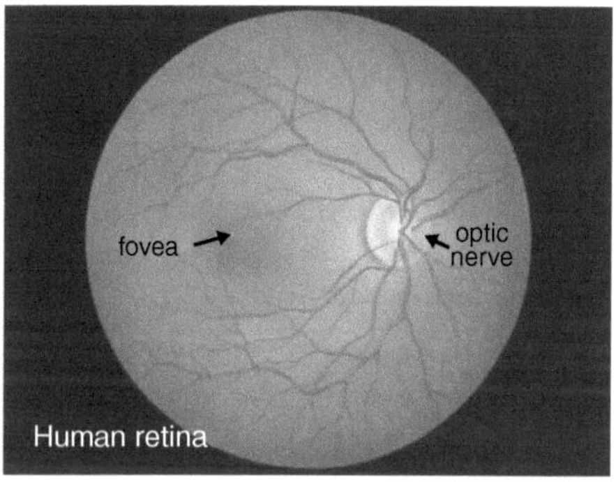

FIGURE 2.7 **(See color insert following page 164.)** Retinal injuries.

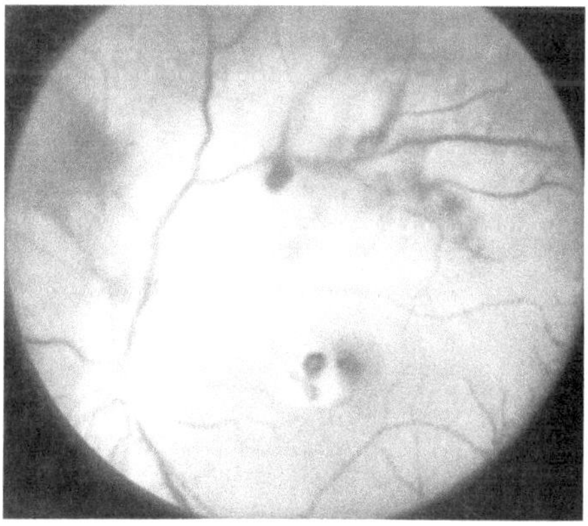

FIGURE 2.8 **(See color insert following page 164.)** Retinal injuries.

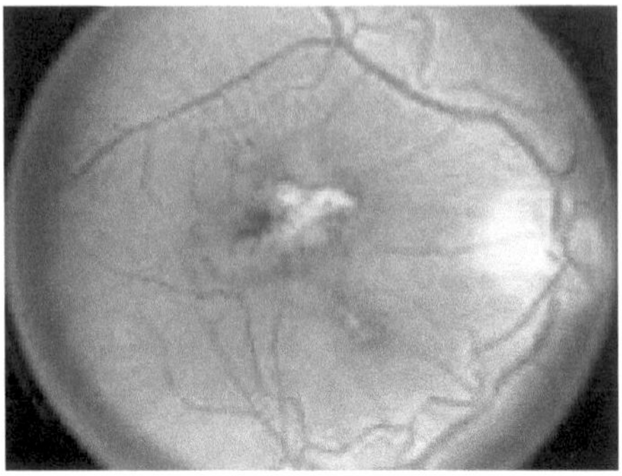

FIGURE 2.9 **(See color insert following page 164.)** Severe fovea and macula damage from NIR exposure.

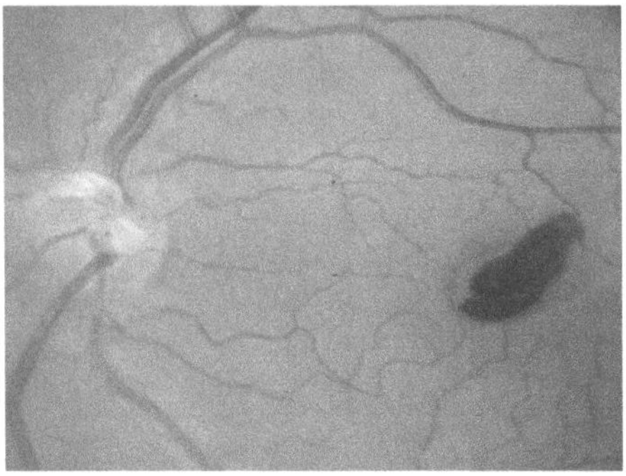

FIGURE 2.10 **(See color insert following page 164.)** Retinal injuries.

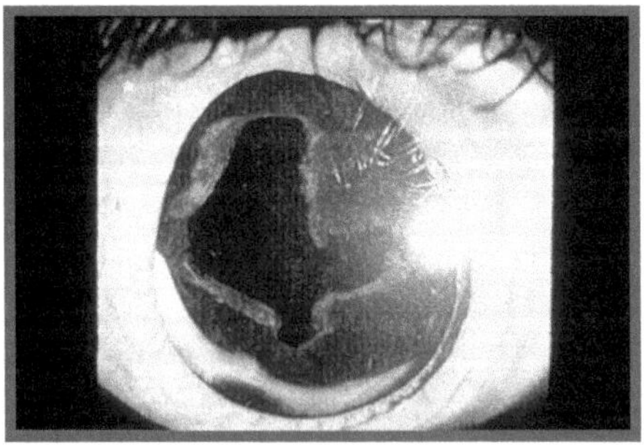

FIGURE 2.11 **(See color insert following page 164.)** Corneal burn.

Melanosomes are also strong absorbers, but their volume fraction in the epidermis may be quite low, perhaps several percent. Again, the local interaction of light with the melanosomes is strong, but the melanosome contribution to the average absorption coefficient may modestly affect light transport.

2.6.1 Scattering Cellular Structures

Photons are most strongly scattered by structures whose size matches the photon wavelength. Scattering of light by structures on the same size scale as the photon wavelength is described by the Mie theory. Scattering of light by structures much smaller than the photon wavelength is called Rayleigh scattering. Following are some structures that scatter light.

2.6.1.1 Mitochondria

Mitochondria are intracellular organelles about 1 μm in length (variable) that are composed of many folded internal lipid membranes. The basic lipid bilayer membrane is about 9 nm in width. The refractive index mismatch between the lipid and the surrounding aqueous medium causes strong scattering of light. Folding of lipid membranes presents larger lipid structures that affect longer wavelengths of light. The density of lipid–water interfaces within the mitochondria make them especially strong scatterers of light.

2.6.1.2 Collagen Fibers, Fibrils, and Fibril Periodicity

Collagen fibers (about 2 to 3 μm in diameter) are composed of bundles of smaller collagen fibrils about 0.3 μm in diameter (variable). Rayleigh scattering from collagen fibers dominates scattering in the infrared wavelength range. On the ultrastructural level, fibrils are composed of entwined tropocollagen molecules. The fibrils present a banded pattern of striations with 70-nm periodicity resulting from the staggered alignment of the tropocollagen molecules, which each have an electron-dense head group that appears dark in electron micrographs. The periodic fluctuations in refractive index on this ultrastructural level appear to contribute a Rayleigh scattering component that dominates the visible and UV wavelength ranges.

3 The Laser Safety Officer: The Key to Laser Safety

3.1 THE EFFECTIVE LSO GETS OUT AND WALKS LASER USE SPACE

This chapter describes a number of laser safety officer (LSO) duties and how to accomplish them. For an LSO to do the job well, he or she must want the position and be allowed the time the job takes. Let us more clearly define the position of the LSO.

Where class 3B, class 4 and nonvisible-beam class 3R products are used, there is the need to appoint an LSO. It may also be appropriate to involve an LSO in risk assessment for the use of class 1M and class 2M laser products.

In the research setting laser safety cannot be achieved and maintained without an LSO. In addition, the LSO must have clear authority to perform his or her role. Without authority and management support, no safety program can survive.

The level of training required by this individual will depend on the nature of the laser application. If the situation is static, that is, the laser application remains the same and is operated day by day, then the LSO may only need to recognize when the risk assessment is no longer valid, either because of changes to the laser application, because the application has entered a different part of the life cycle, or because something has gone wrong. The need for immediate assistance may depend on the critical nature of the application. If you are involved with a 24/7 process, then expert help may need to be on hand all of the time. If time were not critical, then it would be possible to rely on external assistance.

In some countries, general safety legislation requires the appointment of people with the necessary safety expertise, but no legal requirement to have someone with the title laser safety officer. This is why defining the role is important. Collecting an LSO badge will do little to ensure that laser safety is effectively managed unless the appointee knows what he or she is supposed to do and is competent to carry out the task. Using the LSO title may introduce problems in some cultures, and it may be more appropriate to appoint a laser safety manager, recognizing the more general need for the individual to be a good manager, rather than relying on the historical and perhaps defense-oriented officer term.

Laser Identification					Laser Specifications								Direct eye Exposure			Diffuse eye Exposure			Skin exposure
ID #	Type	Make & Model	Comments	Class	Wave-length (nm)	Mode	Beam Size (mm)	Diver-gence (mrad)	Power CW (W)	Pulse Energy (J)	Pulse Length (ns)	Pulse Rate (Hz)	Time (s)	MPE (mW/cm^2)	Min. OD	Time (s)	NOHD (m)	Min. OD @ 0.5 m	MPE for 10 s (mW/cm^2)
1	Nd:YAG	C	-	3b	1062	cw	~2	> 0.35	1	-	-	-	10	5.0		600			1000
2a	ND:YLF	A	-	4	1062	Pulse	1	> 0.35	.2	4.	20	10	10			600			
2b	Diode array	D	--	4	810	Pulse	3	200	-	41	20	20	10			600			
x	HeNe, diode	Various	Several	2-3b	400-700	cw	~1	> 0.42	≤ 15	-	-	-	0.25	2.5	≤ 1.2	600	< 0.1	0	200

SPECIFIC COMMENTS:

WAVELENGTH: at which the laser is operated or capable of operating; **UV** <400 nm, **VIS** 400 to <700 nm, **IR** ≥700 nm, (**near-IR** ≥700 to <1400 nm, **far-IR** ≥1400 nm).

LASER SPECS: typically listed for the smallest accessible beam size, highest power or pulse energy, shortest pulse length, and highest rep-rate.

EXPOSURE TIME: MPEs depend on the length of exposure. Use the actual pulse duration for single pulses; use the following (or greater) for CW or rep-rated pulses:

direct eye exposure: **UV** - 10-30,000 s (i. e. 8-hr work day) (depends on expected exposure time and assumes 2 successive days exposure); **VIS** - 0.25 s (i. e. blink response time) or at least 1/Hz; **all IR** - 10 s

diffuse eye exposure: **UV** - 600-30,000 s (i. e. 8-hr work day) (depends on expected exposure time); **VIS or near-IR** - 600 s; **far-IR** - 10 s

skin exposure: **all wavelengths** - 10 s.

MPE: **M**aximum **P**ermissible **E**xposure for unintentional, intrabeam (direct) exposures for the listed duration – typically in mW/cm^2 for CW or rep-rated (≥ 1 Hz) beams and mJ/cm^2 for single pulses at < 1 Hz. Purposeful direct viewing is not permitted unless authorized specifically in an OSP.

OD: minimum **O**ptical **D**ensity eyewear (at the designated wavelength) for full protection to MPE levels, typically at a distance of 0.5 m from a source. Optically aided viewing with telescopes, microscopes, cameras, etc. may require higher OD. **Note**: "alignment eyewear" for **visible beams 400-700 nm** may be used with an OD reduced by as much as 1.2 than specified in the "Min. OD" column (OD 1.2 is the equivalent to reduce a 15-mW HeNe to MPE level). Employ caution to avoid direct and stray beams. Since there is no aversion response to diffuse light, do not go below the OD level specified in the diffuse "Min. OD @ 0.5 m" column without LSO or OSP-documented approval.

DIFFUSE EXPOSURES: based on 100% Lambertian reflection at normal incidence from a non-specular surface at a nominal arm-length distance of 0.5 m.

NOHD: **N**ominal **O**cular **H**azard **D**istance beyond which laser viewing is safe without eyewear (listed for fiber output and occasionally for unaided viewing of diffuse beams if warranted).

FIGURE 3.1 Laser inventory form.

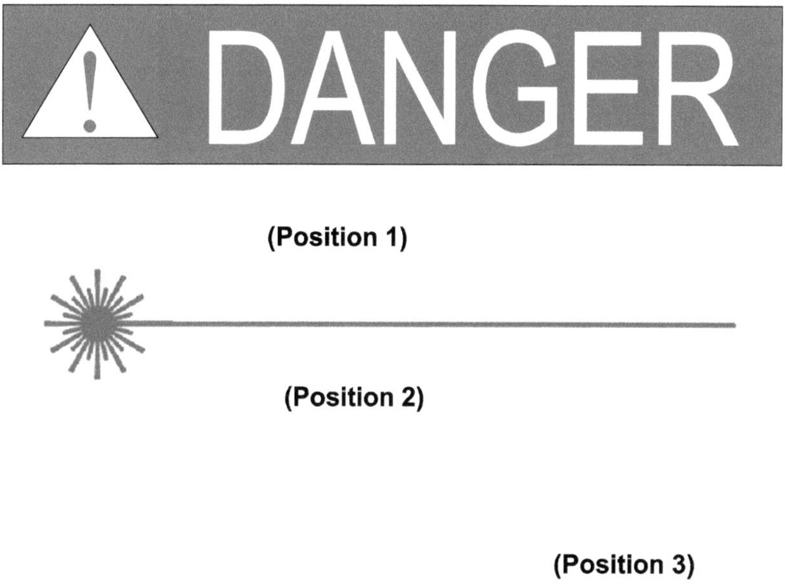

FIGURE 3.2 (See color insert following page 164.) ANST style warning sign, Class 3B and/or Class 4.

In some establishments it may be appropriate to have a number of LSOs, each responsible for an area or application. In order to ensure consistency and the opportunity to learn from incidents, it would then be appropriate to appoint one of the LSOs as the senior LSO or some similar title. There should be an opportunity for the LSOs to meet periodically. The senior LSO may be part of the central safety committee but at least should have the opportunity to communicate effectively with the committee.

Before an LSO or senior LSO is appointed, consideration should be given to the responsibilities, power, and accountabilities of the position. Will the LSO give advice to managers or supervisors in the work area or will he or she have the power to change things? In some places where we have given advice, the LSOs have been given the power to stop laser work if they consider that laser safety is not adequately managed, with the full support of the senior executives in the organizations.

Where the laser application is relatively static or there is a need for additional expertise beyond what could reasonably be expected from the on-site LSO, then it may be appropriate to appoint an external expert. There is no internationally agreed-upon title for this expert, but laser protection adviser and laser safety adviser have been used. The term *adviser* recognizes that, as an external person or body, they cannot have executive power over the organization. They can give advice and, if they feel the advice is not being taken up, either terminate their contract or, in exceptional cases, notify the regulatory authority of their concerns. When appointing an external organization or person to act as an adviser, it is important to clarify the role you wish them to undertake.

3.2 LASER SAFETY COMMITTEE

If you have several LSOs, they can meet periodically. Even with a single LSO, it may be worth considering a laser safety committee. This would allow principal laser users to meet and discuss common issues. It would also allow the LSO to inform users of any changes to legislation and standards. The frequency of the meetings will depend on the complexity of the laser applications and how far along the road to successful laser-safety management the organization is. It is likely that at the start of laser use, it will be necessary to meet at least monthly and perhaps weekly. As the laser application matures, meetings every three or six months may be adequate.

The committee membership needs to contain a representative of management, safety (LSO), and laser users. The laser safety programs at many R&D institutions have been successful without a laser safety committee. In these cases communication between the LSO and users is always active. At times a committee may be formed for a specific short-term goal, such as updating the laser chapter of the safety manual or for an incident investigation.

3.3 INVENTORY

Knowing what types of lasers and wavelengths are in use is critical to developing control measures. In the case of an inventory trap, the name of the laser does not indicate what wavelength it is being used. An Nd:YAG laser is capable of many wavelengths, not just the traditional 1064 nm, but also 532 nm and 266 nm.

How should you obtain inventory data?

1. Consult with your property management group (this might be the best approach and least time intensive).
2. Have the LSO be part of the approval chain or at least get copies of approved purchases.
3. Request an inventory from each department.
4. Perform on-site audits.
5. Have a summer student or intern perform on-site inventory counts.

For many years the most important task anyone involved with laser safety did was count the lasers. While we have moved on to appreciate that just knowing how many lasers you have will not protect you from laser incidents, inventory may still have a part to play in managing laser safety.

In the days when all lasers were either very large, extremely expensive, or both, it was not too difficult to identify where they all were. With modern lasers getting smaller and smaller, and also costing less, we may reach a point where just counting the lasers would be a full-time task. Imagine being in a research laboratory with a bin of 10,000 laser diode chips. Keeping track of them all is not an easy task, nor is it warranted.

We need to keep the purpose of the inventory in mind. Is it for stock control, for security, or for safety management? We will only consider an inventory for safety management purposes, but you may find that this can meet all three requirements. The level of the inventory will depend on the level at which we are managing laser safety. The corporate safety office of a large organization may be interested in the types of lasers and their applications, whereas a single laboratory may need a lot of detail, probably including the following:

1. Laser type
2. Model number
3. Serial number
4. Laser class during normal use
5. Laser class during servicing
6. Date of installation
7. Location

Other information will be useful for managing laser safety and will form part of the information required in Section 1 of the laser-safety management program:

1. Laser medium
2. Means of excitation
3. Initial beam diameter
4. Beam divergence

For some laser products, this information will need to be repeated for each individual laser. To complete the information and to make the record useful for an audit, it is recommended that a brief outline of the laser application be added to the inventory. It is also helpful to include photographs.

No matter how tightly any organization tries to implement their laser safety policy, it can still fall into the trap of inadvertently acquiring a laser product. This could happen because a product includes a laser as an "extra." One example of this is an infrared thermometer. Such products tend to have a sighting laser to show the user what surface the infrared is being detected from. If you just specified a thermometer, you may not have expected a laser with it. Also, a number of infrared remote controls for use with computer presentations have lasers fitted in them.

It is also possible that researchers could order a piece of equipment to carry out a specific task, completely ignorant of the process. As far as they are concerned they are buying a scientific tool. Some equipment may contain one or a number of lasers (i.e. gene squencer, flow cyctometer). It could be argued that the supplier should have informed the researcher before the equipment was ordered, but realistically, that will not always happen. To be fair, such equipment normally presents no risk from the laser beam during normal use or even during user maintenance operations. What happens, though, when the service engineer arrives, removes all of the covers, overrides the interlocks, and starts shining laser beams all around the laboratory? Sometimes that is the first time the user realizes there is a laser in the equipment.

A handy tool for laser inventory is to include a list of the lasers that make up or are found in each standard operating procedure (SOP). The inventory matrix in Figure 3.1 not only lists the laser, but also allows the LSO to perform a safety eyewear analysis of each. A sample form is shown at the end of this chapter.

3.4 TRAINING

While the LSO need not perform user training he must assure that it is accomplished (see Chapter 7 for a full discussion). The way most people tackle laser safety training is to look around for available courses and make a decision based on cost, location, or scheduling convenience. Sometimes the course meets their needs perfectly and sometimes they come away dissatisfied, but more likely they only realize they do not have the necessary tools to move their laser-safety management forward when they try to get things done. This is not to suggest that all laser safety courses are bad. It is more a case of trying to ensure that you get the training you need for whatever role you are intending to undertake. To use the industry jargon, you need to carry out a training needs analysis. In order to do this you need to be able to step back and think through what the task is. Unfortunately, this can be difficult because you may find that you do not know enough without going through a training course to specify what the task will be.

A formal training course may not be the best option. Tutorial-type sessions are available where an external consultant will spend time analyzing the training needs. The LSO may be able to gain enough information during that tutorial to meet his or her own needs, especially if the situation is relatively static and external assistance is available.

Training for laser users tends to be in two parts: the training to undertake the task plus an appropriate level of laser safety training. The appropriate level is important here. Not everyone has an in-depth interest in laser physics and many switch off their attention the moment electrons and energy states are mentioned. Forcing them to sit through several hours of theory would be a complete waste of time. However, there are practical issues that do need to be covered and, of course, these may go beyond any risks presented by the laser beam.

When an organization has a group of people who are going to work with the laser, the most effective laser safety training can be provided around the equipment. It will be reassuring for the workforce to hear how the laser-safety issues in their particular work areas are being addressed. The laser-safety management program can be a useful tool to help with this.

Whatever training everyone receives, a record should be maintained. It is also useful to record whether the effectiveness of any training has been assessed, and how. This could be formal tests or examinations at the end of a course, but it may also be implementation of the material after the course. Again, the production of an applicable laser-safety management program is very effective at demonstrating how competent someone is.

3.5 SIGNS

Signs can be a very visible form of administrative control. The trouble is that they soon become "wallpaper." They are an important aspect of laser-safety management because they are very easy to audit, especially by regulatory authorities. (Quick hint: see Chapter 5, Laser safety tools.)

There are national requirements for safety signs in areas where some laser products are used, as well as requirements under laser safety standards for equipment to be labeled. They are only effective if people notice them, read them, and take some action based on what they have read. Here the term *read* is used in its widest sense since may signs will include some form of pictogram.

Some federal regulations require that you display particular signs. However, these signs should only be displayed when they are relevant. This can be achieved by either covering signs when they are not relevant or turning them over. Signs falling into this category could include a warning about the laser and the environment, that is, whether it is a laser controlled area, some kind of prohibition sign, such as "authorized persons only," and perhaps a mandatory "laser safety eyewear must be worn" sign. It may be appropriate to display information about the laser products contained within the room, but usually this is not important.

How do you deal with beam alignment? For many laser applications this process is carried out by service engineers; in others it is carried out by the user, especially in research environments. One of the most effective ways of displaying a sign is to get a piece of flip-chart paper and put it on the door with an explanatory, clear, written message such as "Laser Alignment in Progress until 3 pm today [insert today's date]. Please do not enter." This type of sign works for two reasons. First, it is big and new. This means there is a higher probability that people will notice it. Second, it is time bounded; that is, you have given a finish time. If it looks as if the work is going to go beyond the stated time, then it is better to alter the time than to put up a new notice. If this notice is always displayed on the door and the researcher is known to be in the coffee lounge or on a week's vacation, then the respect for the notice will be lost.

3.6 WARNING LIGHTS

Warning lights can be an improvement on signs, but the message must be clear and the lights must be located where they can be seen (see Chapter 5 for help). Fitting illuminated panels above doorways is not effective unless the lights flash. There has been a reluctance to fit lights to the sides of, or on, doors because they often present a risk of being bumped by passersby — perhaps even resulting in injuries. However, modern ultra-thin display panels are becoming cost-effective alternatives. They can display warning signs and any other safety information. They can be made to flash if required and usually can be mounted in any appropriate position.

It is important that any light be labeled so that all who need to know are aware of the meaning.

TABLE 3.1
(See color insert following page 164.) Specific Instructions for the Formatting of Laser Area Warning Signs

Safety Alert Symbol & Signal Word	Position 3 Laser Class *right justify, 18-pt*	Position 1 (bold words are ANSI format) *center justify, 18-pt; also see Note below*	Position 2 *left justify, 18-pt*	Reason for use, Comments	Safety symbol
Colors to be used	Use **bold black Helvetica upper & lower case**; background color is white or yellow as shown				
⚠ DANGER Red & white	**Class 4 laser**	**Invisible and Visible Laser Radiation — Avoid Eye or Skin Exposure to Direct or Scattered Radiation**	Information about access, OD requirements, OSP #, etc.; list lasers, OD, wavelengths (or pulse, PRF, output, etc.)	When highest class laser is Class 4	Red
⚠ DANGER Red & white	**Class 3b laser**	**Invisible and Visible Laser Radiation — Avoid Direct Exposure to Beam**	"	When highest class laser is Class 3b	Red
⚠ DANGER Red & white	**Class 3a laser**	**Invisible and Visible Laser Radiation — Avoid Direct Eye Exposure**	"	When highest class laser is Class 3a and above MPE	Red
⚠ CAUTION Yellow & black	**Class 3a laser** *(this sign is seldom used - see comments)*	**Visible Laser Light — Do Not Stare into Beam or View Directly with Optical Instruments**	"	When highest class laser is *visible* Class 3a *and* below MPE for 0.25 s *Note: yellow background*	Black
⚠ CAUTION Yellow & black	**Class 2 laser** *(Class 2a has been eliminated)*	**Visible Laser Light — Do Not Stare into Beam**	"	When highest class laser is Class 2 (all Class 2 lasers are visible) *Note: yellow background*	Black
CAUTION Yellow & black; no alert symbol	N/A	Describe the nature of hazards other than personal injury	Information about potential property damage	For advice on property damage only *Note: yellow background*	Black symbols representative of hazard
⚠ WARNING Orange & white	N/A	Describe the nature of hazards (e.g. controls for servicing ion lasers; unattended laser operation, etc.); provide access instructions or direct to other detailed instructions	Contact information in case of emergency	For advice on potential personal injury if not avoided	Black
NOTICE Blue & white; no alert symbol	Date and name of person posting if a temporary notice.	Describe the nature of the policy or temporary condition (this may supplement DANGER or CAUTION signs but *not* replace them)	Actions to be taken, if necessary, such as before entering	Policy statement; interlocks or status panel not needed; posting outside a temporary laser controlled area; etc.	Black (use symbol if laser related)

3.7 AUDITS

Without the performance of audits, an organization will never have assurances of laser safety. The purpose of the laser safety audit is not to check up on users, but to determine if policies and controls are being followed. If the policies and controls are not being followed, an audit can determine if the neglect is willful or if policies are out of step with the users' needs. In addition, visiting user sites to conduct a laser safety audit allows the LSO to see the space and find out what activities are happening at the institution. One of the values of seeing the space is to determine if the space is right for the work going on. Is there sufficient clearance around optical tables? Are there windows that should be covered? Are there temperature problems in the area? The LSO will never be aware of these issues if he or she never leaves the office.

ANSI Z136.1 Safe Use of Lasers references the LSO's duty of performing audits in several sections: "The LSO shall be responsible for hazards evaluation of laser work areas." "The LSO shall ensure that the safety features of the laser installation facilities and laser equipment are audited periodically to assure proper operation." "The LSO will survey by inspection, as considered necessary, all areas where laser equipment is used."

Therefore, it is safe to say that audits for laser safety compliance are expected to be conducted for facilities using class 3B and class 4 lasers. The composition, frequency, and rigor of that audit rests in the hands of the LSO. A common practice is for institutions to develop laser audit checklists or survey forms. In many institutions, a sole LSO or a number of deputy LSOs perform these audits. Some institutions request users to perform a self-assessment audit.

Many items on the common audit list and its findings are subjective because they are based on the experience and interest of the LSO or auditor in particular items on the checklist. Beam block usage is an example; to one set of eyes, it might be completely adequate, but to another, inadequate. Possibly, experimental considerations prevent additional beam block placement.

Following is a rationale for some of the audit items and a sample audit form.

Interlock log: SOPs that require interlock checks need to be current. Current means one check per quarter (no more than 90 days between checks is the goal). These checks are generally operational performance checks. For complex systems, for the tester is required to follow a written procedure and note problems. The preference is a written procedure for all labs to follow. This ensures consistency between checks regardless of who in the lab performs the check. If problems are noted, follow-up action and documentation of resolution are required.

Alignment procedure: At a minimum, there should be a section in the SOP binder that gives general laser alignment guidance. Whenever possible, laser use-specific alignment procedures should be developed for the different laser activities. System start-up procedures could go in this section.

3.7.1 Posting and Labeling

Hazard communication poster: Laser hazards need to be on a hazard communication poster. Check the poster to see that it represents all hazards in the room, not just your work.

Laser sign: Laser use areas with class 3B or class 4 lasers are required to be posted with a laser warning sign. The sign should accurately convey the wavelengths in use and any laser protective eyewear requirements. The sign needs to be on all accessible entrances to the laser use area.

Emergency contact: Many lab doors have emergency contact information posted. It must be readable and accurate.

3.7.2 Beam Enclosures

The goal is to contain the laser beam and any stray radiation to the optical table or intended use area. Enclosures that confine the beam are one of the best methods to accomplish this. This means individual portions of the laser beam can be contained as in a beam tube, or the optical set-up can be contained by means of a barrier around the entire table or portions of it. This barrier can be several inches higher than the intended beam path, open or closed at the top, or with panels several feet high enclosing the entire table.

Total enclosure: This is the preferred but not always possible method. Panels can be labeled with interlocked or non-interlocked warning labels.

Totally open: While not preferred, in some cases this may be the only workable option. In such a case, use of properly placed beam blocks is critical to safety. A check for stray reflections is required after each alignment or beam manipulation.

Combination: In some cases beams will not be totally open or totally enclosed, but a combination of both. A combination approach is acceptable and realistic.

Perimeter guard: A guard must be of sufficient height above the intended beam height to prevent a likely stray reflection from rising above it.

Beam tubes: For open distances between optics over 2 feet, it is recommended that beam tubes be mounted. It is preferred but not required that the tube be made of a material opaque to the laser radiation. Keeping hands out of the beam is the major goal.

3.7.3 Open Path Walkway

At times the laser beam may need to pass from one optical table to another across an open walkway space. In such cases the level of controls can vary depending on the hazard presented by such a beam. If the beam is below MPE, the user may choose to use administrative means, that is, signs.

Permanent beam tube: This is the preferred control, but it may present emergency or traffic control problems.

Removable method: To avoid the problems presented by a permanent beam tube, a removal tube may be the best solution. In such cases, a control must be in place to prevent the laser radiation from crossing the open space without the tube or awareness of users in the lab. Some approaches: removable tubes; tubes, bars, or swing arms; chains across the area; and swing gates.

3.7.4 Protective Eyewear

Laser protective eyewear is a critical part of laser safety for the individual. Chiefly, it relies on the user to wear the proper eyewear and take care not to abuse the eyewear. All laser users need to know they have an obligation to make sure everyone in the laser lab wears the proper eyewear when a laser radiation hazard is present.

Full protection: This type is designed so that the optical density of the eyewear will absorb all the laser radiation from a direct hit for a period of up to 10 sec. Intrabeam or direct viewing of the laser beam is strictly forbidden. Do not stare into laser with remaining eye.

Alignment: Use of alignment eyewear is allowed for visualization of visible beams for alignment activities. The NIF LSO grants approval of such eyewear.

Labeled: The labeling required is the optical density (OD) and wavelengths the eyewear is designed to provide protection from. Labeling on some common styles of eyewear can wear off. All eyewear needs to be labeled and readable; otherwise it must be removed. Labeling can be self adhered.

Quantity: The quantity on hand must be sufficient for the expected number of daily users and anticipated visitors. Visitors should be limited to full protection eyewear only.

Condition: Laser eyewear must be in good condition, free from scratches, abrasions, and burns in critical vision areas. The LSO needs to be contacted to determine if the eyewear still provides the level of protection required.

Correct OD: The OD on eyewear must meet the limits set forth in the laser table for the laser application.

Prescription age: Because of the cost of prescription laser eyewear, the user may use a pair with a prescription several years old. A consult with health services is required to determine if a new set of eyewear is required.

Storage: Eyewear must be stored in a manner that preserves its condition. Storage can be outside the laser use area or inside. Each has advantages and disadvantages.

Holder: The storage of laser protective eyewear will have a direct effect on its useful lifespan. The practice of eyewear being thrown in a drawer or left on tables at the end of the day is unacceptable.

3.7.5 Beam Containment

Beam blocks: These should be made out of a noncombustible material for the power output expected to strike the block. They must not transmit the wavelength in use. Cardboard may be suitable for some applications, while metal will be required for others. The block should not be reflective for the wavelengths being used. All active beam blocks must be secured to the optical table (unless foot print stops tip over and are approved by the LSO). The size of the blocks must be sufficient to block the beam diameter and possibly a misaligned beam. A label on the block indicating that it is a beam block and not to be moved is recommended but not required. Such labeling is considered a good practice to help locate any beam blocks that might be misplaced or knocked over.

Path to door blocked: There should not be a direct path from the laser use area to a door leading outside the laser use area. Blockage of this path can be provided by a curtain, a partition, or a barrier on the laser table.

3.7.6 Fiber Optics

Container for sharps: If it cuts and slices fibers, a container is required for sharp ends, following standard sharps protocols.

Fiber ends labeled: Near or where the beam exits the fiber, a label is required to make staff aware of this hazard point (unless alternate controls have been approved).

Conduit labeled: Conduits carrying laser radiation levels above class 1 need to be labeled at least every 3 m and at points where they enter or exit a wall.

3.7.7 Housekeeping

Laser work surfaces: The area on the optical table encompassing and directly adjacent to the beam path needs to be free of all nonessential reflective sources. This includes optics, tools, foil, and storage containers. This does not include established alternate beam paths for related experiments.

Related work surfaces/adequate storage space: Space is always of a premium in any laboratory; the more organized the space is, the safer the work area will be. Divisions and programs should make resources available to individual labs to aid in this goal. Users have a dual responsibility here: first to remove unused equipment, either to surplus or storage outside the lab, and most importantly to maintain an ongoing effort to organize and put away supplies.

Chemical storage: Chemicals present several hazards to the user as an ignition and combustion source to an irradiant. Their proper storage and use is critical to personnel safety, particularly when laser dyes are in use. In choosing secondary containers, keep in mind that while plastic trays are cheap and easy to obtain, they do not offer the resistance to fire that metal trays do, nor do they give off toxic chemicals when burning.

High storage: Adherence to seismic guidelines needs to be checked on shelves and on the tops of racks.

Trip hazards: A number of commercial devices are available to protect cords and hoses from tears and prevent people from tripping over them. The application of these devices needs to be reviewed during the audit.

Emergency lighting: The auditor should make sure emergency lighting is present, working, and located in adequate locations.

Auditor ____________ SOP # ____________ Audit Date: ____________

Type of Audit: ☐ Annual ☐ New ☐ Amend ☐ Self-Assessment Other

Facility Name: ____________ Building: ____________ Rooms: ____________

Principal Investigator: ____________

Room Contact During Audit: ____________

Classes of lasers in this experiment/rooms: ☐ Class 4 ☐ Class 3b ☐ Class 3a ☐ Class 2 ☐ Class 1 embedded

Posted Documentation and Security Measures

1. Access door interlocks & status panel functional: ☐ Key ☐ Code ☐ Code Comments: ____________ ☐ Y ☐ N ☐ NA
2. Access door signs current format, emergency contact current: Comments: ____________ ☐ Y ☐ N ☐ NA
3. Posting on ancillary doors: ____________ ☐ Y ☐ N ☐ NA
4. Current SOP available: ____________ ☐ Y ☐ N ☐ NA
5. Eyewear requirements posted: ____________ ☐ Y ☐ N ☐ NA
6. Interlock check sheet available & current: ____________ ☐ Y ☐ N ☐ NA
7. Alignment procedure (class 3b & 4): ____________ ☐ Y ☐ N ☐ NA
8. Interlock check procedures available for complicated systems: ____________ ☐ Y ☐ N ☐ NA

Laser unit safety controls:

9. Laser classification labels present on commercial units: ____________ ☐ Y ☐ N ☐ NA
10. Protective housings in place: ____________ ☐ Y ☐ N ☐ NA
11. Beam shutters interlocked & functioning as per interlock check sheet: ____________ ☐ Y ☐ N ☐ NA
12. Interlock bypass functioning: ____________ ☐ Y ☐ N ☐ NA

Engineering and Administrative Laser Safety Controls

13. Lasers & optics secured to table: ____________ ☐ Y ☐ N ☐ NA
14. Beam properly contained (not a hazard to persons sitting or standing): ____________ ☐ Y ☐ N ☐ NA
15. Beams enclosed where available: ____________ ☐ Y ☐ N ☐ NA
16. Noncombustible, nonspecular beam barriers in place: ____________ ☐ Y ☐ N ☐ NA
17. Adequate controls where beams leave tables or leave enclosures: ____________ ☐ Y ☐ N ☐ NA

FIGURE 3.3 Sample Audit Form.

18.	All beams attenuated or low power when practical: ____	☐ Y	☐ N	☐ NA
19.	Windows/door openings covered: ____	☐ Y	☐ N	☐ NA
20.	Beams blocked from open by-passed doors: ____	☐ Y	☐ N	☐ NA
21.	Nonessential reflective materials out of beam paths & surroundings: ____	☐ Y	☐ N	☐ NA
22.	Administrative controls employed, barriers demarcated: ____	☐ Y	☐ N	☐ NA
23.	Upward directed beams are labeled: ____	☐ Y	☐ N	☐ NA
Other Safety Measures:				
24.	Laser eye exams by all personnel (3b and 4 lasers): ____	☐ Y	☐ N	☐ NA
25.	Proper eyewear available for all personnel, ODs OK: ____	☐ Y	☐ N	☐ NA
26.	Proper storage of eyewear, where? ____	☐ Y	☐ N	☐ NA
27.	Proper skin protection available & employed: ____	☐ Y	☐ N	☐ NA
28.	Are all current users on training matrix? ____	☐ Y	☐ N	☐ NA
29.	Collecting optics used (microscopes, binoculars, telescopes): ____	☐ Y	☐ N	☐ NA
Non-beam Hazards:				
30.	High voltage hazards minimized: ____	☐ Y	☐ N	☐ NA
31.	Optical tables bonded to building ground: ____	☐ Y	☐ N	☐ NA
32.	Optical tables seismically secured (if no, how many?): ____	☐ Y	☐ N	☐ NA
33.	Housekeeping fire hazards minimized: ____	☐ Y	☐ N	☐ NA
34.	Good housekeeping on optical tables: ____	☐ Y	☐ N	☐ NA
35.	Fiber optic use: ____	☐ Y	☐ N	☐ NA
36.	Container for fiber sharps: ____	☐ Y	☐ N	☐ NA
37.	Fiber ends/connectors labeled: ____	☐ Y	☐ N	☐ NA
38.	Fiber conduit labeled: ____	☐ Y	☐ N	☐ NA
39.	Other non-beam hazards minimized: ____	☐ Y	☐ N	☐ NA

Comments:

Auditor signature ____________________ Date ____________

FIGURE 3.3 *(Continued)*

4 Standard Operating Procedures: A Binding Safety Contract

4.1 INTRODUCTION

A key element of any laser safety program in a research environment is a standard operating procedure, commonly called an SOP. Different institutions may have their own names for this document: operational safety procedures, activity hazard document, work authorization document, and so on. Regardless, the basic sections are the same, even if the format varies. From a laser safety perceptive, the SOP needs to be considered as a contract between the laser users and the laser safety officer (LSO) or safety department. While the focus of this book is laser safety, a truly complete SOP contains a description of all hazards associated with laser work (toxic chemicals, pressure, radiation, electricity, etc.) and their planned mitigation procedures.

The principal investigator (PI) or supervisor must prepare an SOP detailing the methods, responsible individuals, and materials used with the laser and have this approved by the LSO before beginning research. The typical SOP contains the following information:

1. The location, building, and room
2. A description of the work to be performed
3. A description of the laser including the class and the beam characteristics listed on the manufacturer's label
4. The intended use of the device and the type of research
5. An analysis of the potential hazards and the establishment of safety parameters
6. The use of the appropriate personal protective equipment PPE (especially eye protective equipment) suitable to limit laser beam exposure
7. A description of the emergency procedures and contingency plan necessary to deal with accidents and injuries

In addition, the SOP may include:

1. The sign-off on the line management that authorizes the individual to conduct research within the boundaries stated in the SOP

2. A diagram of the work area
3. The time frame during which the document is valid (expiration date)

4.2 SOP FLOW

4.2.1 The PI or Responsible Individual Develops the Draft SOP

Emergency, safety & hazard (ES&H) personnel may assist in drafting the SOP, but the PI has the primary responsibility. An important thought process occurs when individuals must state in black and white what they are going to do and how they are going to do it. Over involvement by ES&H personnel in an attempt to "help" may do more harm than good by removing the PI from the safety thought process. Safety must remain integral to the R&D development process. R&D personnel must be safety conscious at all times.

4.2.2 The SOP Must Have a Complete and Accurate Work Description

A work description sets the boundaries within which the work will be conducted. It is an agreement between the experimenter and all reviewers and approvers regarding what work will be conducted. It can be amended and reviewed if the scope of work needs to change. It must be updated if and when the scope of work changes.

4.2.3 Hazards Must be Clearly and Completely Identified

A checklist is helpful to ensure that nothing is overlooked. Every hazard should have some type of a stated control. All parties (PI, management, ES&H, facility personnel) should concur on the SOP.

4.3 SAMPLE FORMAT

4.3.1 General Information

1. Title
2. Location of work
3. Date of preparation
4. Division
5. Activity supervisor (name of person with authority to designate experiment operators)
6. Principal investigator

4.3.2 Description of Activity

Provide a brief summary of the work and a complete description of activities to be performed. Include details of unique equipment or materials and special application of standard equipment or materials. Include experimental parameters, such as quantities of chemicals used and stored, gas pressures, operating temperatures, voltages, current, and so on. Describe energy sources (electrical, pressure, gravitational, chemical, thermal, etc.). Describe the temporal aspects of the work (continuous or periodic, around the clock, etc.). Describe the need for and basis for any engineering notes appended to the SOP.

4.3.3 Identification of Hazards

A comprehensive hazard identification and hazard analysis must be performed to identify the potential hazards associated with the activity. The relationship of the identified hazards and their potential effects on the experimenter, proximate personnel, the public, property, and the environment should be analyzed. The preferred hazard identification and analysis is specific and not generic (e.g., "electrical shock" is insufficient; "contact with live 220V AC power from the bus bar if front panel is removed or if interlock is bypassed" is more appropriate). General identification can work if, in addition, SOP specific hazards are called out.

4.3.4 Mitigation of Hazards

Describe in detail the controls necessary to mitigate the identified hazards. Controls may include, but are not limited to, engineering controls (interlocks, special ventilation systems, fire detection and suppression systems), personal protection and other safety equipment (gloves, face shields, aprons, eyewash/safety shower, etc.), and administrative controls (lockout/tagout and other special operating procedures, distance or time constraints, special training, etc.).

4.3.5 Hazardous Material Handling

Describe all hazardous materials involved. Include quantities used, as well as storage, handling, segregation of incompatibles, and labeling requirements. Describe the process for communicating information on material safety data sheets (MSDSs). Describe how the MSDSs are maintained.

4.3.6 Hazardous Waste

Identify hazardous waste generated by the activity and disposal criteria. Determine the waste generated and the control and authorizations necessary to deal with it. This may include special training and permit and storage protocols. An additional factor is to identify any waste stream permits that may be required.

4.3.7 Emergency Procedures

Provide procedures designed to respond to emergencies that may be associated with the activities described above. Issues to consider may include:

1. Potential to ignite or fuel a fire
2. Special fire extinguishers or unique fire prevention measures
3. Special first aid or medical response criteria unique to the identified hazards
4. Notification of the responding fire department regarding unique fire safety or medical-response issues associated with the work
5. Discussion of unique medical implications with responding medical teams
6. Special emergency shutdown procedures
7. Impact of a sudden power failure
8. Emergency escape plan
9. Supplemental emergency communications needs
10. Hazardous material spill or release response (control, containment, and training) procedures
11. Inventory of spill or release response supplies
12. Emergency equipment inspection protocol and schedules

4.3.8 Maintenance

Consider inspection and maintenance procedures for materials, equipment, and components identified in Section 4.3.4 as critical parts or components that can lead to a catastrophic failure. An example would be testing room access interlocks and eye wash showers.

4.3.9 Authorized Users and Training

Provide a list of the authorized users and a list of required training courses prerequisite to becoming an authorized user. Include a matrix that documents users, formal EH&S training course requirements, training completion dates, refresher training dates, on-the-job training, and so on. Include a user signature line (with date) for each user to sign, confirming their understanding of and conformance with the SOP.

4.4 FOLLOW-UP TO THE SOP

An on-site prestart review should be conducted before final approval to begin work. The PI, management, and ES&H should be represented. The scope of work, hazards, and controls are again discussed to ensure that nothing has changed. Controls for hazards should be verified such as:

1. Correct laser eyewear present
2. Beam tubes in place

3. Interlocks installed, functioning, and tested
4. Proper warning devices such as signs and warning lights in place

The SOP should be reviewed periodically (annually at least) to ensure there have not been significant changes to:

1. The work description
2. Hazard levels
3. Implementation of hazard controls
4. Authorized personnel

Two key elements for a laser-based SOP are laser alignment procedures and laser controls. Following is a sample of laser alignment guidance, followed by a list of useful boilerplate laser control items that should be included in all laser-based SOPs.

4.4.1 Preparation for Alignment

1. To reduce accidental reflections, remove watches, rings, dangling badges, necklaces, and reflective jewelry before any alignment activities begin. Use of nonreflective tools should be considered.
2. Access to the room or area is limited to authorized personnel only.
3. Consider having at least one other person present to help with the alignment.
4. All equipment and materials needed are present prior to beginning the alignment.
5. All unnecessary equipment, tools, and combustible materials (if the risk of fire exists) have been removed to minimize the possibility of stray reflections and nonbeam accidents.
6. Persons conducting the alignment have been authorized by the responsible individual.
7. A notice sign is posted at entrances when temporary laser control areas are set up and when unusual conditions warrant additional hazard information being available to personnel wishing to enter the area.

4.4.2 Alignment Methods to be Used for This Laser

4.4.2.1 Alignment Considerations

1. Whoever moves or places an optical component on an optical table (or in a beam path) is responsible for identifying and terminating each and every stray beam coming from that component (meaning reflections, diffuse or secular).
2. There must be no intentional intrabeam viewing with the eye.
3. Coaxial low-power lasers should be used when practical for alignment of the primary beam.

4. Reduce beam power with ND filters, beam splitters, or dumps or by reducing power at the power supply. Whenever practical, avoid the use of high-power settings during alignment.
5. Laser protective eyewear must be worn at all times during the alignment, within the parameters and notes specified in the accompanying laser table.
6. Beam blocks must be secured (and labeled if possible).
7. Have beam paths at a safe height, generally below level when standing or sitting.
8. The LSO has authorized eyewear with reduced optical density (OD) to allow the beam spot to be seen. Measures shall be taken and documented to ensure that no stray hazardous specular reflections are present before the lower-OD eyewear is worn. Maximum-OD eyewear, as listed in the laser table, is to be worn again once alignment is complete. The reduced-OD eyewear is labeled as alignment eyewear and is stored in a different location than the standard laser eyewear for this operation.
9. Skin protection should be worn on the face, hands, and arms when aligning at ultraviolet (UV) wavelengths.
10. The beam is enclosed as much as practical. The shutter is closed as much as practical during course adjustments. Optics and optics mounts are secured to the table as much as practical. Beam stops are secured to the table or optics mounts.
11. Areas where the beam leaves the horizontal plane must be labeled.
12. Any stray or unused beams are terminated.
13. Invisible beams are viewed with infrared (IR)/UV cards, business cards, card stock, craft paper, viewers, index cards, thermal fax paper, or Polaroid film or by a similar technique. Operators are aware that such materials may produce specular reflections or may smoke or burn.
14. Pulsed lasers are aligned by firing single pulses when practical.
15. Intrabeam viewing is not allowed unless specifically evaluated and approved by the LSO. Intrabeam viewing is to be avoided by using cameras or fluorescent devices.

4.4.2.2 Alignment Conclusion

Normal laser hazard controls are to be restored when the alignment is completed. Controls include replacing all enclosures, covers, beam blocks, and barriers and checking affected interlocks for proper operation.

4.5 LASER CONTROL BOILERPLATES

The object of laser control boilerplates is not to repeat the requirements of a laser chapter, but rather to set forth a number of good practices to make sure all users read them as part of the SOP.

4.5.1 General Laser Controls

1. All laser operations must be conducted in accordance with the institution's EH&S laser safety chapter.
2. All operating personnel must have had a laser eye exam prior to working with class 3b or class 4 lasers.
3. All laser-operating personnel must have completed approved laser safety training prior to working with class 3b or class 4 lasers as well as an orientation to the laser experiment by the PI, including an explanation of the beam path, reflective surfaces, and hazards associated with the work.
4. The PI must be responsible for ensuring that all personnel, visitors, and students with access to the laser use area have a clear understanding of the controls associated with laser operation.
5. Laser operators must use appropriate OD laser protective eyewear.
6. This eyewear must be stored in such a manner as to protect its physical integrity. There must be sufficient laser protective eyewear on hand for users and expected visitors.
7. Laser areas must be kept as clean as possible. Reflective objects such as tools, optics, screws, and so on must be kept away from laser beams.
8. Appropriate laser warning signs, approved by the LSO, must be posted on all entrances to the laser use area.
9. In laser use areas that have room access interlocks, the interlocks must be tested quarterly. The results must be noted by the PI or designee on the interlock check sheet kept in the interlock check section of the SOP ES&H binder.
10. Laser beams should not be directed toward the entrances of any laser use area. In such cases, beam confinement or barriers must be in place to protect anyone entering the laser use area from direct exposure to laser radiation.
11. Whenever possible (within experimental considerations), appropriate enclosures, barriers, beam blocks, or beam tubes must be applied to contain laser radiation above the threshold that could cause eye or skin damage.
12. Any beam paths between tables or between laser tables and targets where an open walkway exists must have a control in place to block those in the room from crossing the pathway (e.g., hinged tubes, beam tubes, retractable tapes, or chains).
13. Whoever manipulates or moves optics is responsible for checking for stray reflections. When found, those reflections must be contained to the optical or experimental table(s), even if they are below an eye hazard level.
14. The LSO must be notified if any lasers are added or any experimental changes are made to an approved laser set-up that impact safety. At such times, the LSO must perform a hazard evaluation.

4.6 SAMPLE LASER SOP1

Laser:	Date:
Department:	Location:

4.6.1 Laser Safety Contacts

Primary laser operator:____________________	Phone:__________________
Laser safety officer:____________________	Phone:__________________
Maintenance/repair:____________________	Phone:__________________
Medical emergencies:____________________	Phone:__________________

4.6.2 Laser Description

1. Location of laser or laser system (site, building, room)
2. Diagram of area layout with beam path, including locations of emergency shut-offs, fire extinguishers, protective equipment (eyewear), and barriers (attachment)
3. Description of each laser, including classification, lasing medium, and beam characteristics (divergence, aperture diameter, pulse length, repetition rate, and maximum output, as applicable)
4. Purpose and application of beams

4.6.3 Laser Safety Program

Clearly outline each category below:

1. Responsibilities of the laser operators
2. Security or warning system activation or key control
3. Safety procedures
4. Laser training requirements
5. Personnel protective equipment requirements

4.6.4 Operating Procedures

1. Initial preparation of laboratory environment for normal operation (key position, outside status indicator on, interlock activated, warning sign posted, personnel protective equipment available, etc.)
2. Target area preparation
3. Special procedures (alignment procedures, safety tests, maintenance tests, etc.)
4. Operating procedures (power settings, Q-switch mode, pulse rate, etc.)
5. Shutdown procedures

4.6.5 Control Measures

Laser/Laser System Controls

Check if applicable	Control	Comments
[ ]	Entryway (door) Interlocks or controls	
[ ]	Laser enclosure interlocks	
[ ]	Laser housing interlocks	
[ ]	Emergency stop/panic button	
[ ]	Master switch (operated by key or code)	
[ ]	Laser secured to base	
[ ]	Beam stops/beam attenuators	
[ ]	Protective barriers	
[ ]	Warning signs	
[ ]	Reference to equipment manual	
[ ]	Extra eyewear available	

Comments:

Hazards and Controls

Check if applicable	Hazard	Controls
[ ]	Unenclosed beam/ Access to direct or scattered radiation	
[ ]	Laser at eye level of person sitting or standing	
[ ]	Ultraviolet radiation/blue light exposure	
[ ]	Reflective material in beam path	
[ ]	Hazardous materials/waste(dyes, solvents, other)	
[ ]	Fumes/vapors	
[ ]	Electrical	
[ ]	Capacitors	
[ ]	Compressed gases	
[ ]	Fire	
[ ]	Housekeeping	
[ ]	Trip hazard	

Comments:

4.6.6 Personnel Protection Equipment

A. Eyewear

Laser Eyewear					
For this laser....			...Wear this eyewear		
Acquisition #	Type	Wavelength(s) (nm)	Wavelength(s) Attenuated (nm)	Optical Density(OD)	Remarks
Example	Nd:YAG	1064,532	1064,532	5+	UVEX

B. Other Protective Equipment Required within Nominal Hazard Zone

Item Location Usage Condition

____________ ____________ ____________

____________ ____________ ____________

4.6.7 Operator Review

I have read and understood this procedure and its contents and agree to follow this procedure each time I use the laser or laser system.

Name (print) Signature Date

____________ ____________ ____________

____________ ____________ ____________

____________ ____________ ____________

____________ ____________ ____________

4.7 LASER AND LASER SYSTEM SOP GUIDELINES

I. Guidelines for SOPs

Written SOPs must address normal operations as well as beam alignment for each laser or laser system. In addition, they should include nonbeam hazard management and servicing of the laser or laser system.

Keep SOPs brief to increase their probability of use.

It may be helpful to utilize the laser safety manual and the laser training form in developing your SOP.

The Office of Environmental Health & Safety will be happy to review and comment on your draft SOP. The LSO may be reached at ____________

II. Beam alignment (concerns to be addressed in the SOP)

A. **Security:** Secure the lab to avoid distractions by marking the door with an appropriate sign or marker such as "NOTICE: Laser Work In Progress. Do Not Enter. Eye Protection Required."

B. **Beam Characteristics:** Is the beam visible or invisible? Is special equipment needed to view the beam? If the beam is pulsed, is it possible to fire one pulse at a time to limit exposure time and hazards?

C. **Beam Viewing:** Intrabeam viewing is prohibited on Tulane campuses, and a remote viewing device should be used if intrabeam viewing is required for beam alignment. Only diffuse reflections may be viewed directly. Always use the lowest beam power possible that will still allow viewing of an image with proper protective eyewear.

D. **Personal Protective Equipment (PPE):** Use laser protective eyewear with a low enough OD to allow viewing of the diffuse reflection. Use appropriate body shields (gloves, lab coat, UV face shield) to protect against UV beam scatter.

E. **Personnel:** It is ideal to have two people involved in the alignment process at all times. If this is not possible, at least let someone know where you will be and when you plan to have the alignment process completed. Regular check-ins are advised.

F. **Replace Beam Controls:** Ensure that all beam locks, enclosures, and beam barriers are replaced when the alignment is complete.

G. **Remove Door Signs:** Verify that the "NOTICE: Laser Work In Progress. Do Not Enter. Eye Protection Required." sign is removed from the room entrance and that the regular ANSI laser warning is still in place.

III. Normal operation of the laser (concerns to be addressed in the SOP)

A. **Security:** Doors opening to areas where lasers are in use must be secure. A closed door does not constitute security. Doors must be rendered unopenable by using standard door locks or by activating door interlocks.

B. **Personal Protective Equipment:** Have the appropriate safety equipment on hand. Specify what is needed and its specific use.

C. **Start-Up Procedures**: Insert key, turn on water, turn on power supply, close shutter, activate laser, and so on as specific to your laboratory.

D. **Experimental Procedure:** Specific to your laboratory.

E. **Emergency Procedure:** Note the location of the emergency shut-off mechanism, emergency procedure posting, fire extinguisher, safety shower, emergency eyewash station, and so on.

F. **Shutdown Procedure:** Specific to your laboratory.

G. **Storage:** Remove laser activation key and deactivate interlocks. Store these devices in a secure location.

IV. Nonbeam hazards to address (concerns to be addressed in the SOP)

A. **Electrical Hazards:** Only properly trained personnel may work with high-voltage systems. It is ideal to have two people involved in the alignment process at all times. If this is not possible, at least let someone know where you will be and when you plan to have the alignment process completed. Regular check-ins are advised. Laboratory staff may be trained in CPR as a precaution.

B. **Fire Protection:** Attention should be given to preventing fires and explosions. Flammable solvents are frequently used as laser dyes and to clean optical parts. Fire extinguishers should be well marked and staff should be versed in their proper classification and use.

C. **Compressed Gases:** Staff should be trained in the safe management of cylinders and the associated hazards of the compressed gases being used.

D. **Toxicity of Lasing Media**

5 Practical Tools for Laser Safety and Traps to Avoid

Users and LSOs need a toolbox full of toys.

5.1 INTRODUCTION

A number of products and solutions exist in the marketplace to help people achieve laser safety. Some are for the laser safety officer's (LSO's) use, some are for the user, and some would be classified as engineering controls, while others fall into the administrative control camp. The goal of this chapter is to make you aware of these tools. Such tools include laser hazard evaluation software, warning signs and labels, beam alignment and viewing aids, power meters, laser eyewear samples, beam termination and laser radiation containment materials (curtains and barriers), digital cameras, and reference sources.

5.2 THE LASER SAFETY "TOOL BOX" WISH LIST

As the authority with jurisdiction over the laser safety program, the LSO is responsible for ensuring "the knowledgeable evaluation and control of laser hazards." The LSO must do this within the confines of a limited safety budget. The organization is rare that does not have to continually justify its expenditures. This applies in particular to expenditures for safety programs, including laser safety. While effective safety departments save money by facilitating injury and illness prevention, they are seldom viewed as revenue generators.

5.2.1 Laser Hazard Evaluation Software

An important LSO responsibility is recommending or approving protective equipment, including laser-attenuating eyewear and windows. In order to properly recommend or select such eyewear or windows, the LSO must determine the required optical density (OD). To do this, the LSO has several options: rely on a vendor to determine the correct OD, use the guidance provided in the American National Standards Institute (ANSI) Z136.1 standard to manually calculate the OD, or use commercial software to perform the necessary calculations. Depending on the variety of laser wavelengths used at a facility, software may be the optimal choice. Using software has certain advantages, particularly if one is not routinely doing calculations or has many of them to do. By using default values for some of the OD calculation parameters (for example, the correction factor and recommended exposure duration), the software simplifies the task of determining OD values.

The authors would be remiss if they did not note that all of the hazard evaluation software programs they have reviewed have some flaws or quirks. In order to confidently use these software programs, the LSO should have a feel for what the correct OD value should be given the particular laser output parameters and the potential exposure condition, for example, intrabeam versus diffuse reflection. While all the calculations can be done manually, hazard evaluation software does them quickly and provides a detailed report that can be appended to a standard operating procedure (SOP) or other laser safety–related documentation.

Generally, vendors sell licenses to use hazard evaluation software rather than selling it outright. The purchase of a corporate license may be less expensive than purchasing separate licenses for multiple locations within one's organization.

5.2.2 Warning Signs and Labels

Approving the wording of signs and equipment labels is another LSO responsibility. To ensure that signs and labels are both accurate and properly posted, the LSO may wish to obtain and provide laser users with such signage and labels. Typically, the LSO has three choices regarding sign and label acquisition: buy the signs and labels from a commercial source, use software developed specifically for making in-house signs or labels, or use a generic software program (Claris Draw®, MacDraw®, PowerPoint®, Freelance Graphics®, etc.) to generate in-house signs and labels.

If signs and labels are a limited need or rarely needed, purchasing commercial signs and labels may be the preferred option. When there is a routine requirement for signs and labels, the latter two choices are often quicker less expensive. Using these software options, one can easy customize both signs and labels; location-specific information, such as LSO contact number, user contact numbers, and OD requirements, can be easily incorporated. When deciding how to obtain laser safety–related signs and labels, the LSO should consider the different types of signs and labels that may be needed. Depending on the scope of the laser safety program, the LSO may need laser certification labels, manufacturer identification labels, aperture labels, warning logotype labels and signs, inventory labels, and protective housing warning labels. Making one's own signs and labels using software becomes more attractive when one also takes into account nonlaser labeling and signage needs such as emergency shutdown button sign-age, lockout and tagout labels, high voltage warning labels, and collateral radiation warning signs.

5.2.3 Digital Camera

A digital camera can be a great asset to any laser program. It can be invaluable for documenting laser-related accidents or incidents. Investigative reports are enhanced,

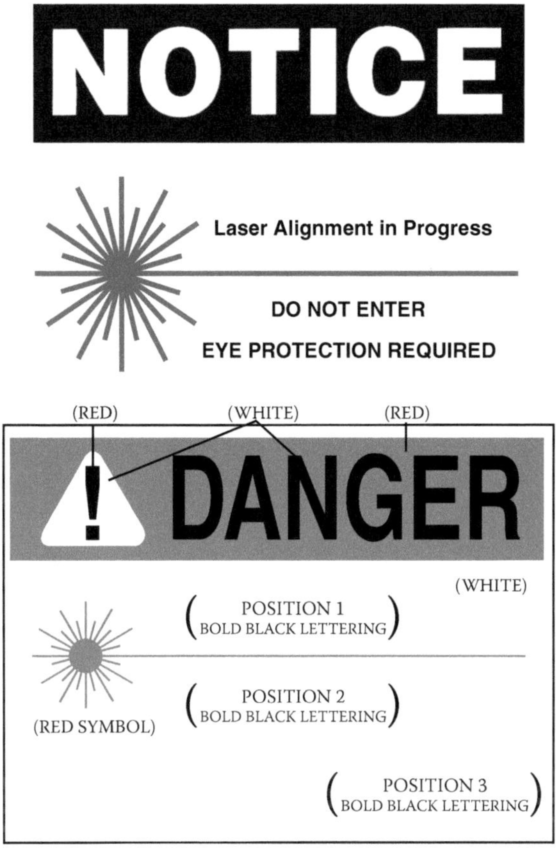

FIGURE 5.1 **(See color insert following page 164.)** (a) Laser danger sign as of 2000; (b) notice sign: used for repair or alignment notification.

and the accident scene is more readily understood. Digital images can be used to clarify SOPs, audit reports, and control measure descriptions and aid in describing safety concerns. Such images can also be important adjuncts to any regulator reports, such as Center for Devices and Radiological Health (CDRH) reports.

As with other devices involving information storage technology, the cost of digital cameras continues to decrease. When new models are introduced with enhanced features, sales outlets for these cameras drop the prices on older inventory items. One need not have a camera with an inordinate amount of megapixels to obtain clear digital images for training as well as for incident prevention and investigation efforts.

5.2.4 Laser Eyewear Samples

Failure of laser eyewear to fit well or comfortably is a primary reason given for not wearing laser protective eyewear. By having samples of frames or complete laser eyewear available, the LSO can help ensure that the eyewear purchased will be used. Laser users can see whether or not the protective eyewear feels good on their faces and if its visible light transmission (VLT) is sufficiently high to allow the safe performance of critical tasks. Ordering unsuitable eyewear is avoided, money is saved, and the turnaround time for obtaining useful eyewear is reduced.

5.2.5 Resource Material

It is important that the LSO keep abreast of new laser technology developments that may have a future impact on the laser use environment in which he or she has laser safety responsibilities. Despite the plethora of laser safety–related information on the World Wide Web, magazines and books are still valuable information sources for the LSO. While it is important to have various written reference sources, whether in hard copy or on the Web, there is no substitute for talking with a consultant or peers when encountering an unfamiliar laser safety issue. Information exchange and networking is seen by nearly all safety and health professionals as a value added activity. Participation in professional associations allows one to meet individuals with laser safety challenges similar to one's own.

5.3 TRANSITION ZONE BETWEEN THE LSO AND LASER SYSTEM USERS

5.3.1 Beam Alignment and Viewing Aids

As the ANSI Z136.1-2000 standard points out, "laser incident reports have repeatedly shown that an ocular hazard may exist during beam alignment procedures." Aids for beam alignment or viewing include low-power lasers (class 2 and visible class 3R), sensor cards or screens, and viewer or image converters. The use of lower-power laser beams' paths can simulate the beam paths of higher-power lasers while minimizing the potential exposure hazard. Converters and sensor cards can be invaluable to both the LSO and laser users in determining the adequacy of efforts to properly block beams and their stray reflections.

5.3.2 Viewers

In general, infrared (IR)-wavelength alignment aids are generally more available than their ultraviolet (UV)-wavelength counterparts. IR viewers enable their users to locate invisible beams in the near-IR (NIR) portion of the spectrum, generally up to 1300 nm, although IR viewers can be found that extend the IR viewing range out to 1500 or 1700 nm (Figure 5.2 and Figure 5.3). While less common than their IR viewer counterparts, UV-wavelength viewers are also available.

5.3.2.1 How Do They Work?

IR and UV viewers are image converters rather than image intensifiers. The principle is based on a photocathode image converter. It translates the IR power

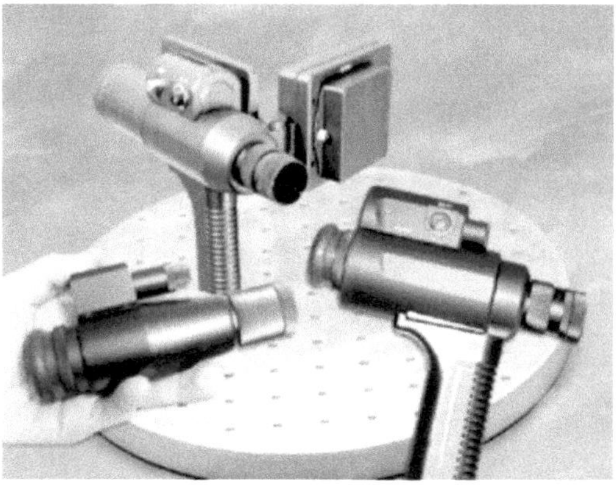

FIGURE 5.2 Assorted infrared viewers.

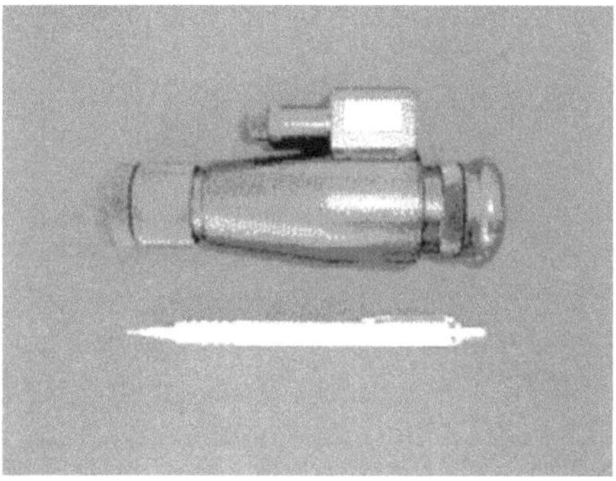

FIGURE 5.3 IR viewer (size comparison).

distribution into an image visible through the eyepiece. An extremely low-level (picowatts of IR) is sufficient to produce an image.

5.3.3 Sensor Cards

Similar in purpose to viewers, sensor cards can also be used to locate invisible beams, either UV or IR (Figure 5.4). These cards and converters permit invisible UV and IR beams and their reflections to be "seen." The phosphors in these devices are designed so that one can see the up-converted or down-converted invisible wavelengths. One should select sensor cards with care since they respond to different invisible wavelengths and minimum power density (irradiance) thresholds. Some sensor cards generate specular reflections, and appropriate laser eye protection should be worn as a precautionary measure. Sensor cards are available in a variety of sizes and shapes, and some of them require preactivation using UV or visible radiation.

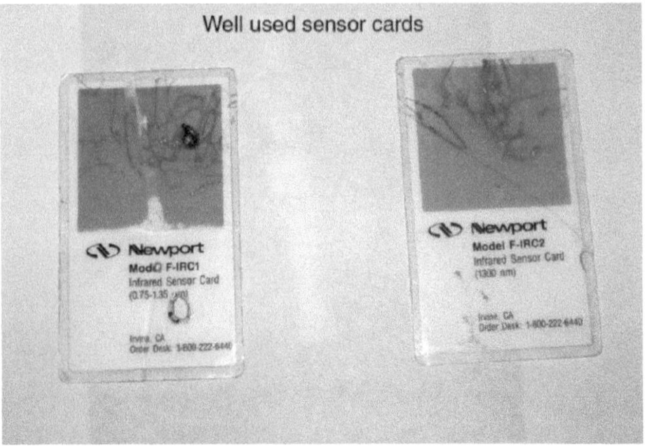

FIGURE 5.4 Well-used IR sensor card.

5.3.3.1 How Do They Work?

The cards contain a special sensitive area coated with a chemical phosphor layer, which emits clearly visible light when illuminated by NIR, IR, and some UV sources. IR sensor cards work on a principle known as *electron trapping*, where phosphor-based compounds are employed to absorb and "trap" incoming light energy from a short wavelength and release that stored light in the form of visible light upon stimulation from a longer IR wavelength. The visible result is a localized glow, which is relative in intensity to the amount of stored light and IR power levels exciting the active area. Many cards have the sensor area encapsulated between durable clear plastic layers, allowing for easy handling.

5.3.4 Optical Glass Sensors

While not as common as IR sensor cards, optical glass sensors do have applications. They are constructed of optically clear glass with specific phosphors adhered to one side of the glass substrate. IR radiation emitting from a laser or LED source will be visualized from either side of the sensor screen.

5.3.5 Beam Blocks and Dumps

Beam blocks, backstops, traps, dumps, and stops provide a means to terminate unneeded laser energy and keep beams confined to a particular area, such as an optical table. These can be commercially obtained or manufactured in-house. An example of an in-house beam block would be L-shaped 1/8 to 1/4 inch thick, 8 to 14 inches high, and 2 to 12 inches wide, made of anodized aluminum. The anodization process should be of a beaded pattern, rather than a flat shiny black surface. The diffuse surface reduces the chance of specular reflections. The base of the L would have, depending on the width, holes drilled into it, allowing it to be secured to an optical table. The beam blocks can be located behind the optics or turning mirrors. Beam dumps are generally used as part of a system to capture a split-off beam

5.3.6 Beam Containment Enclosures

Enclosures around laser setups or systems are a great aid to laser safety. These can be composed of interlocked or noninterlocked panels. Once again, commercial firms manufacturer enclosures to user specifications or they can be build in house. Usually, they are a metal frame with opaque panels.

5.3.7 Curtains

Laser curtain barriers are either used in a fixed mode capacity or a temporary set-up. The LSO at one time or another may face the issues surrounding the need for a temporary control area barrier. While such a barrier is usually needed during equipment service and maintenance, it can also be useful during laser alignment in an open area or an area not designed for open beam hazards. A fixed mode of use would be when one is trying to separate the laser use area from a larger space. These curtains are similar in appearance to many welding curtains; they are designed to withstand either direct or, more commonly, diffusely scattered laser beams. Barriers must be opaque to the laser wavelengths in use, cannot be combustible, and must be designed to withstand the intensity of the laser beam that may strike them. Barriers can be passive or active. Most are passive. Commercial laser curtains can be a rather expensive; at times more cost-effective alternatives are available. Commercial laser curtains are certified by the manufacturer to withstand laser irradiation in general at the level of 100 to 300 W/cm^2 for beam sizes of 5 mm or less when the exposure time is 100 sec or less. Higher-irradiance models are also on the market. Laser curtains can be purchased on portable frames to provide a temporary containment when it is needed.

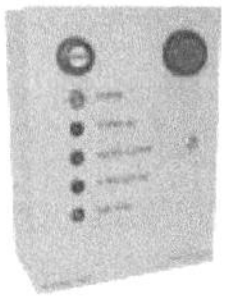

FIGURE 5.5 Rockwell Laser Industries interlock system.

5.3.8 Active Barriers

The active guard or enclosure incorporates a dedicated sensor into the passive guard to detect the presence of an errant beam on the enclosure wall. The guard has only to maintain its integrity for a time sufficient for the sensor to detect the beam and feed a signal back to shut down the laser. The sensor may be intrinsic or extrinsic to the guard. Active guarding systems were developed for use in conjunction with CO_2 lasers and Nd:YAG laser fiber optic delivery systems.

An example of an active guarding system for use with high-power CO_2 lasers is the hot spot detector (HSD) developed by the Safety and Reliability Directorate (Culcheth), Atomic Energy Authority (Culham), and SIRA. The HSD is an errant beam detector; its function is to detect potential hot spots on the walls on the enclosure and to feed back a signal to shut down the laser system. It was considered that the detector could make it possible to relax constraints on enclosures to resist penetration and use more lightweight enclosures. After several trials of the system to determine response time compared to burn-through time, detection ranges, and so on, the following enclosure strategies were proposed for use in conjunction with the HSD.

For powers up to 1 kW (continuous wave [CW]), 1.5- to 2.0-mm painted steel panels with 6-mm thick polycarbonate windows provide a safety factor of 1:10 in terms of detection to burn-through time for beams of 10 mm; for powers >10 kW, 1.6-mm painted steel with 6-mm polycarbonate windows provides only a safety factor of 1:4 (1:5 deemed acceptable). Double skin panels filled with sand or water would improve the safety factor.

Because of the variation of burn-through time with beam polarization, it was suggested that 100-mm thick thermal building bricks would provide the best solution. Clearly, the use of such fortress style enclosures defeats the initial object of the HSD in offering an inflexible solution. Furthermore, this active approach is highly reactive, its entire operation being based on the occurrence of the hazardous event.

5.3.9 Room Access Interlocks

In some circumstances access to a laser use area may require an interlock system. The purpose of these systems is to keep unauthorized personnel out. The majority of laser use facilities build their own interlock access systems. A limited number of commercial laser access interlock systems are on the market. These systems shutter or deactivate laser systems in use. At times just room access control will achieve a similar level of safety. This includes key card access. Card access, while not affecting the laser, has many advantages. It can restrict access to only those authorized and even monitor training compliance.

5.3.10 Illuminated Warning Signs

Particularly when a class 4 laser is in use, an indicator of that use is required. The most common means of indicating that use is through an illuminated sign

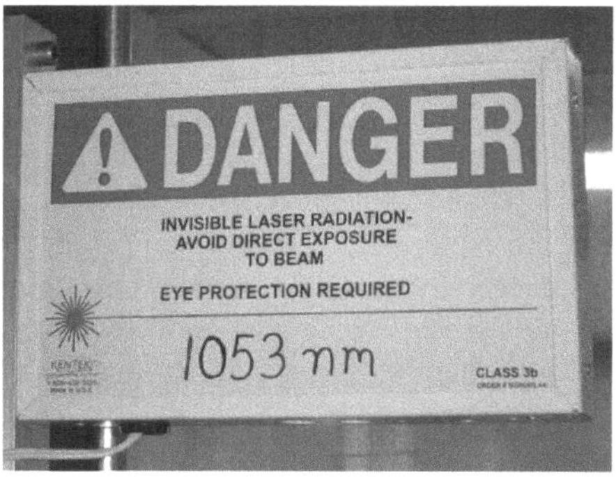

FIGURE 5.6 Illuminated warning sign (can flash or be steady light).

or visual indicator (light) immediately outside the laser use area (Figure 5.6). These devices can be obtained commercially. A word of advice: Such illuminated signs will serve you better if the illumination source is LED based. The extended lifetime of LEDs versus standard light bulbs or torches can justify not incorporating a failsafe mechanism into the warning sign. With a fail-safe mechanism, if the light burns out, the laser will not be able to fire or a shutter will drop in front of the laser beam.

Illuminated signs fall into different styles:

1. Illuminated warning sign that looks like a laser warning sign
2. Illuminated sign with message levels, that is, laser off, laser powered, beam on
3. Illuminated signal with no indication of purpose (red light over door)
4. Illuminated programmable sign (with custom information, such as wavelength information)

The light over the door, in the author's opinion, needs to have an accompanying sign indicating what it means, for example, "light on equals laser on."

5.3.11 Power Meters

The LSO can use laser power meters to determine or confirm the classifications of laser products or systems. For example, the LSO can use a power meter to determine if the output of a helium neon laser labeled class 3b has degraded enough to allow its reclassification as class 3a. Such reclassification allows the relaxation of laser safety–related control measures for that laser's use. The LSO may also use a power meter to self-certify laser barriers and attenuating windows. Loaning power meters to laser users may also help develop good rapport between those users and the LSO. The majority of power meter use is designed for user applications.

All power meters contain the four basic elements. Light striking the detector results in a current flow through the detector. The amplitude of the current is proportional to the total optical power striking the detector for power within the detector design range. This current is amplified and used to drive the meter read-out device. This may be an analog or a digital display. It is the detector's role to convert the incoming optical signal into an electrical signal, which is then amplified and recorded. Most power meters are modular. They typically consist of two parts: a main control unit that features a read-out display, signal-processing electronics, and an interface for transferring data to a computer, and a detector head that is placed in the path of the light beam to be measured. Many manufacturers sell a wide variety of detector heads designed for different power levels. These simply plug into the main unit and can greatly enhance the flexibility of a power meter.

Power meters can be divided into two types, each of which detects power by a different method. The first type uses thermal (thermopile) detectors, while the second type relies on semiconductor photodiodes. A special kind of pyroelectric thermal detector is also available for measuring pulse energies rather than power.

If you are looking for a general-purpose power meter for measuring CW laser powers of more than a few milliwatts, a thermopile-based solution is probably the most appropriate. Thermopile detectors offer a very wide, flat spectral response with a high damage threshold. They are a popular choice for those working with more powerful lasers such as Nd:YAG, Ti:sapphire, carbon dioxide, excimer, and argon ion. Detector heads for taking measurements from around 0.5 mW up to 10 kW are commercially available with a spectral range from 200 nm to 11 μm.

For very sensitive power measurements at a wavelength of less than 2 μm, a semiconductor photodiode is the best choice. These can measure powers as small as a fraction of a picowatt and are often available with heads that are specially designed for accepting optical fiber connectors for those working in the telecommunications field. Semiconductor-based meters are also available for making pulse measurements such as pulse energy and peak power.

If you need to measure light from a strongly diverging source such as a laser diode or the end of a bare optical fiber, consider an integrating sphere. If you are looking for a low-cost, convenient solution for quick but less accurate measurements, battery-powered hand-held semiconductors and thermopile probes are now available. However, these cannot transfer data to a computer and are not as precise as more sophisticated models.

When selecting a power meter as a consumer purchase, it is important to know what you want and to find a product that meets your particular applications. In doing this you will find yourself with the correct product, not one with features you will never use or need. So, determine what you need by asking yourself a number of questions:

Do I need to measure the power of CW beams or the peak power or energy of individual pulses? Most meters are designed for one task, although some are compatible with heads that allow both. Thermopile-based systems are often useful for measuring the power of CW lasers or the average power of pulsed lasers above a few milliwatts. Photodiode-based systems can also make both pulsed-power and CW measurements but are often limited to low powers. Both photodiodes and thermopiles can calculate or infer the energy per pulse, while pyroelectric detectors can directly measure pulse energy.

What is the maximum and minimum detectable power I need, and is a probe with the appropriate performance available? If you need to measure very low power levels (below 1 mW), then a photodiode-based solution is often best. Sensitive, low-noise photodiode heads that can measure subpicowatt power levels are now available. By contrast, if you are checking power levels in the watt or kilowatt regime, a thermopile solution is best.

What response time do I need? Thermopile probes often need to be left in the beam for a few seconds. Semiconductor detectors, however, are designed to have a fast response.

Can the meter connect to a wide range of probes in case requirements change? Probes can usually be disconnected from the meter and exchanged with a different model to suit different power ranges or wavelengths. See what is on offer and how much the individual probes cost.

Is the meter's calibration traceable to internationally recognized standards? If the results from the meter are to be trusted, it is vital that it has been properly calibrated. For complete confidence, check to see if the meter is traceable to a standards body such as the U.S. National Institute of Science and Technology (NIST). Also find out at what wavelength the calibration was performed; ideally, you want it to be as close to your operating conditions as possible. A properly calibrated meter will be able to make a measurement with an uncertainty of less than 2 to 3%. Be sure to recalibrate the meter regularly; otherwise it may start to lose its accuracy. Most firms recommend annual recalibration. Also find out what calibration costs are from the manufacturer; this is a hidden cost that may affect how often you can afford to calibrate your equipment.

How large is the meter's detector entrance aperture? Make sure that the diameter of your light beam is not larger than the detector. If you need to measure a strongly diverging beam, then consider using your detector with an integrating sphere, which is guaranteed to collect all the light.

5.3.12 Detectors

Detectors used in laser measurements can generally be characterized as photon detectors and thermal detectors. The photon detector responds to the number of individual photons incident onto the active surface of the detector, whereas the thermal detector responds to total optical power. In the thermal detector this input radiation is absorbed by the detector, producing a temperature increase that then results in a change in some other parameter, such as resistance, contact potential, or polarization. Thermopile detectors, also known as thermal detectors, measure optical power by sensing the heat that is released into an absorber when it is irradiated by a light beam. The detector head is made of two parts: an absorbing front surface and a cooled heat sink. A thermocouple built into the head is used to generate an electrical signal that relates to the difference in temperature between the absorber and the heat sink. This signal is passed to the control unit for conversion into an optical power reading.

Thermopiles are detectors that measure beam power by measuring temperature differences within the detector that are produced by the heating effect of the laser beam. They consist of metal disks connected to heat sinks at their edges. This heat sink may be either air cooled or water cooled. Thermocouples located at the center of the disk and the outer edge are connected in series to produce a voltage that is proportional to the temperature difference from the center of the disk to the edge. During CW power measurements, the thermopile operates in a steady state with a constant heat flow and constant temperature difference. When the power changes, new thermal equilibrium conditions must be reached before the reading will be accurate. Thus, the response time of a typical thermopile is longer than 2 sec. Thermopiles are useful for all wavelengths at power levels above 1 W. Air-cooled detectors are used for powers up to 300 W, and water-cooled models may measure powers of a few kilowatts.

5.3.12.1 Pyroelectric Detectors

This type of thermal detector uses a ferroelectric crystal instead of a thermocouple to sense the temperature difference. Incident photons heat the crystal, causing an electric current to flow. This kind of detector is used for directly measuring the energy of an optical pulse. It cannot be used for measurements of optical power. Pyroelectric materials are nonconductors in which the electrical polarization is a function of the temperature of the material. It consists of a slab of pyroelectric material with electrodes deposited on the surface where the polarization appears. The charge on the electrodes corresponds to the polarization of the material and thus to its temperature. When the temperature changes, current flows through the load resistor. Thus, pyroelectric detectors respond to the change in detector temperature and cannot be used for direct measurement of CW laser power. For CW radiometers, the input beam to the pyroelectric detector is converted to a pulsed input by a light chopper. The system is calibrated to indicate the true CW power of the beam. Pyroelectric radiometers are useful from the blue end of the visible spectrum through the far IR for low-power beams. They are most important in IR regions where other low-light-level detectors cannot be used.

5.3.12.2 Semiconductor Photodiode Detectors

Photodiodes are usually made from silicon, germanium, and indium arsenide and directly convert incoming photons into an electrical signal. This kind of detector offers a fast and sensitive response, but they are highly wavelength dependent.

5.3.12.3 Integrating Sphere Detectors

These are spherical domes that are used to collect and homogenize the light before it enters a detector. An integrating sphere detector ensures that all the power is collected from a strongly divergent source and is a good way to obtain very accurate, reproducible results.

The most common photodiode detector is the PIN photodiode. In this device there is a region of intrinsic semiconductor material between the p and n regions. This intrinsic region increases the width of the junction, which in turn provides for more efficient conversion of photons to charge carriers. These devices can be operated in either a photoconductive or photovoltaic mode. Silicon PIN photodiode detectors are probably the most widely used laser detectors in the visible and NIR regions of the spectrum.

5.4 LASER SAFETY TRAPS

A number of traps or obstacles to laser safety have developed with the growth of laser technology. This section highlight those traps, in the hope that an alert laser user will avoid them. These traps have been the cause of or a contributing factor to many a preventable laser accident.

5.4.1 800-nm Iceberg

To paraphrase Rod Sterling, "You are now entering the iceberg zone." It is well known that only 12 to 20% of an iceberg is above the waterline, meaning it is extremely hard to determine its actual size by visual perception alone. The same is true of light past 700 nms. Our eye detects less than 1% of the available photons at such wavelengths, thus yielding a faint image. Mentally, we equate that faint visualization with a weak or low power source.

Many a researcher has been injured while attempting to view, for example, 751-nm or 810-nm beams they perceived only faintly. In almost all these cases, the user knew the actual beam output, but that fact was masked by the faint equal weak theorem, or iceberg effect. Needless to say, they were not wearing any protective eyewear to look at such a weak beam. NIR laser diodes, Ti:shappire, and Alexandrite lasers are some of the lasers that have been the source of such wavelengths in many of these incidents.

The best way to avoid this trap is awareness, making sure users have been informed of this problem and reminded of it. In addition, alternative means of viewing such as CCD cameras, IR sensor cards, and, as always, laser protective eyewear are good options.

5.4.1.1 Beam Visualization Chorus

If Pink Floyd's *The Wall* were a laser safety song, the most famous line would be, "How do you align if you can't see the beam?" This refrain has been a chorus from laser users since the dawn of laser technology. Of course, it has a true side, but rather than being the mantra that justifies laser alignment without protective eyewear, it should be the signal to call the LSO for input.

The purpose of alignment eyewear is to allow the user to visualize the beam while lowering the intensity of any beam that is transmitted through the user's eyewear to a class 2 level. To address this issue there is a European norm that recommends the OD for alignment eyewear versus the output of lasers used.

This author has often met laser users who proudly show off their laser protective eyewear for visible lasers, but when asked how they align with them on, the answer is silence or a lifting up of the eyewear. So, how do you align safely? One option to consider is the use of laser-alignment eyewear, which provides partial protection and allows visibility. A second is remote viewing by a CCD camera, which, especially with the newer small cameras, can be positioned to view a target, mirror, or other optics. With the addition of motorized mounts, alignment can be made without the users being at risk. The cost of CCD cameras and video monitors has dropped sharply in the past few years, making this a real option. As a third option is the old standby of lowering the laser's output power during alignment by neutral density filters or control of power supply current. Very simply, full power is rarely needed to align a beam path. A fourth option would be using a low-power coaxial beam to show the beam path, generally a class 3A (or 3R). Yet another simple option is fine-tuning the alignment by the use of an iris shutter or a series of irises and holes in a senor card. A simple but under utilized technique is the Hartman plate for telescope alignment. Last, we have fluorescent plates, paper, and sensor cards; even some cross-hair setups can aid in alignment procedures.

TABLE 5.1
European Norm 208 Optical Density recommendations for Alignment Eyewear

Scale #	OD	Max instantaneous power Continuous Wave laser (W)	Maximum energy for pulsed lasers (J)
R1	1–2	0.01	2×10^{-6}
R2	2–3	0.1	2×10^{-5}
R3	3–4	1.0	2×10^{-4}
R4	4–5	10	2×10^{-3}
R5	5–6	100	2×10^{-2}

5.4.2 Reflections: Visible or Invisible

A number of laser incidents can be traced to stray reflections from optical setups. These incidents fall into three categories:

1. A reflection that was present but unknown and not looked for
2. A reflection that was known but thought not to be a problem
3. An instant reflection generated by some action (e.g., moving the power meter into live beam, reflection off a tool)

5.4.2.1 Present but Unknown Reflections

During the set up and alignment of a laser system, it is essential to stop several times to check for reflections leaving the plane of the table. Using an IR or UV viewer or lowering the light level are acceptable approaches.

Consider this example from a September 2004 Department of Energy incident report: While aligning the diagnostics for an ultrafast Ti:Sapphire class 4 laser (800 nm), an experimenter raised his laser safety eyewear to rub his eye to alleviate an irritation due to an existing eye infection. He felt a bright flash and afterwards a light cloudiness in his left eye. Repairs on the laser were completed earlier in the day. In his eagerness to get his experiment underway, the experimenter introduced beam onto the table while he aligned the optics. He rotated one of the polarizing beam splitters. The secondary beam was not considered or accounted for, therefore not blocked or contained. By doing so, an unwanted/undetected beam left the plane of the optical table at an upward 45 degree angle, which subsequently struck his eye.

5.4.2.2 Known Reflections

One of the biggest mistakes one can make is to know of a reflection that leaves the table and not block or contain it because one does not think it is worth the effort or presents a hazard. An experimental setup had an invisible reflection (3000 nm) leaving the table at such a steep angle that it struck a spot 8 feet up on an adjacent wall. Because the reflection left the table at such a steep angle and no one was over 8 feet tall, the decision was made to just disregard it — until someone was being shown how to place the offending optic in place. During the placement of the optic, the reflection traveled up and down the wall and struck the person being instructed, who was standing directly opposite the optic to get a better view of the procedure. Another fact in this incident is that a coaxial HeNe (633 nm) beam ran with the 3000-nm beam. The injured person could have worn eyewear that would have allowed 633-nm visualization but blocked the 3000-nm beam, but no eyewear was worn.

5.4.2.3 Instant Reflections

The technique of moving a power detector head into an active beam is poor practice, no matter how one defends it. This practice is not as uncommon as we in laser safety would like to think. LSOs start examining power meter detector heads for burn marks and burn-off coatings. Jewelry and ID badges are ready-made reflective sources if not removed (in the future all such valuables can be sent to this author for safekeeping). In such cases injury could be prevented if all staff in the laser use area wear laser protective eyewear. Poor housekeeping on the optical table is a set-up for a reach into the beam path or bringing reflective tools, optics, and so on into the beam path.

You may ask, "How can I see visible reflections if I am wearing my eyewear?" This is a valid question because even with alignment eyewear, some visible diffuse reflections may be hard to see. A possible solution is to view the room through a digital or video camera; this way you might be able to image the source area. Other options are to look from a known safe vantage point or view the area with an IR viewer — IR may be leaking through with the visible light.

5.4.3 Housekeeping

Any investment counselor will tell you real estate is a premium asset. The same is true for laser labs. If only our labs were like the TARDIS from the BBC *Dr. Who* TV series, which is larger on the inside than the outside. For those of us who have not solved the space-time dimensional problem, space is a real issue. Even as lasers become smaller, we still find objects to fill all our space. Therefore, many laser labs looks like people's garages. Unfortunately, this clutter is not confined to the space around the optical tables, but is also on optical tables themselves (even vertical optical table set-ups are not immune to clutter). Spare optical mounts, tools, lenses, mirrors, plastic bags, and plastic and cardboard boxes all tend to find a home on optical tables. Rather than being in a staging area separate from beam paths, they are under and next to active beam paths, acting as lures to attract hands into live beams or a reflection source when lifted through the beam.

Some solutions are arranging a staging area on the optical table outside the active beam line; constructing a second, upper, level on the optical table; and organizing cabinets, removing unused items that have been taking up space since 1980. Secondary storage for such items is an option all facilities should consider as well as a housekeeping day once a quarter. Even if you clean up a lab today, things will creep back in over time. Setting designated housekeeping days, just like preventative maintenance days, frees one from the pressure of stopping while project deadlines loom.

5.4.4 Fiber Optics

Say "fiber optics," and the response is like the story of the blind men describing an elephant; the answer all depends on what part they are feeling. Most people respond with a telecommunication response, but a laser user might give a completely safe system response. While both responses are accurate, they do not capture the whole of fiber optics in laser technology today. Laser radiation delivered through fiber optics today (CW to femtosecond pulses) extends to telecommunications, medicine, robotics, manufacturing (welding), and scientific applications.

Safety folks like fiber optics because it contains the laser beam; the bare fiber can be jacketed to provide additional protection. So, how can this frail fiber be a safety trap? The rule of thumb for years was that the divergence of the beam from the end of a fiber is so broad that it was not an eye hazard beyond about 10 cm (4 inches for the metric challenged). Here is where the trap starts. Many fiber applications today include a micro lens at the end of the fiber, creating a collimated beam rather than a quickly diverging one, so the hazard zone can be meters long. Second, the amount of energy being transmitted through fibers has steadily increased, as demonstrated by the development of diodes and diode arrays. Third, most wavelengths applied to fiber work are invisible, and the fiber end is far from the source. This leads to the issue of whether the fiber is active or not when it is found disconnected (warning labels are a good way to solve this problem). Fourth, handling and cutting or splicing fibers presents a sharps and UV hazard. One can see that fiber lasers, while a great asset to laser technology, are not free from problems. Then there is the issue of broken fibers (a number of operating room fires have been started by laser beams escaping from broken fibers).

6 The Laser-Safety Management Program

A path is being set before you, take along a binder

6.1 INTRODUCTION

The laser-safety management program as presented here is a tool that is designed to demonstrate the current status of one's laser safety and to act as a plan for where you need to be and how you are going to get there. While the format presented here is paper based, there is no reason why all the documents should not be stored electronically. However, most people seem to find the paper version easier to use. In an electronic version certain sections could be linked and updated from larger databases such as a training database. In this way as a user completes a course, it would be entered into the program's training matrix.

6.2 BENEFITS OF THE LASER-SAFETY MANAGEMENT PROGRAM

When there are many other things you have to do at work, it is reasonable to question why you have to put more effort into something that could be seen as even more bureaucracy. The approach presented in this chapter is based on over a decade's experience of helping organizations identify their laser safety issues, manage those issues, and prove that they have done so.

Whom are you preparing the program for? Primarily your own organization, but there are different groups of people within the organization who could benefit from a clear statement of your laser safety system, what you are trying to achieve, and how you intend to get there. These include the following:

1. Senior management
2. Safety office
3. Laser management
4. Laser users

The laser-safety management program also has benefits because of the presentation to people external to the organization. These include enforcement officers, but perhaps more importantly, anyone wishing to fund your work. You may also find that the program fits in well with any quality-control system your organization uses.

Perhaps the biggest test for laser-safety management programs has been their implementation in research and university environments, where any form of administrative load is seen to get in the way of valuable research time, particularly for postgraduate students working for their doctorates in a fixed time. However, we have been able to demonstrate that the program is just an extension of good project management. The students can take ownership of their laser-safety management program. Pressure from supervisors can be used to ensure that students thought through the work they wanted to do — mainly practically, but with safety issues included. As the students came to write their theses, they found that they had a chapter on safety management. Think of the impact this has on assessors from industry. It is not unusual to employ a new doctorate to find that he or she has had no safety training and essentially has an extremely poor safety ethic. The students who had used the laser-safety management program were able to demonstrate that they were, at the minimum, safety aware and recognized the positive benefits of this approach.

6.3 FILE STRUCTURE

The approach we have generally used is to use a ring binder file with 13 sections. Consider the binder as a central point for laser safety documentation. With use it may be found that your laser application needs more or fewer than 13 sections, and they may not have the same titles as those suggested here.

We need to decide on the level of the laser-safety management program. At the highest level, we could produce a program for the whole organization. The advantage of this is that a standard could be set across the organization. We could have a program for the site, department, section, laboratory, or individual laser application. As we get closer to the laser application itself, more detail is likely to be included. We also need to exercise some caution. If we have a laser-safety management program for each laser application, and there are a number of applications in the laboratory, they cannot be treated in isolation.

6.4 INSTITUTIONAL BINDER SECTIONS

1. Safety structure
2. Safety committee
3. Standards and regulatory
4. Training program
5. Audits and reports
6. Risk assessments
7. Control measures
8. Incident investigation
9. External liaison
10. Miscellaneous

We suggest starting with the following 13 sections for user binders:

1. Standard operating procedure
2. Training and authorized worker list
3. Interlock checks
4. Alignment procedures
5. Audits and reports
6. Material safety data sheets and URLs
7. Checklists
8. Engineering safety notes
9. Contact list
10. ES&H chapters
11. Lessons learned
12. Emergency procedures
13. Miscellaneous

The following sections are optional:

14. Safety structure
15. Specific procedures
16. Safety committee minutes
17. Standards and regulatory items
18. Risk assessments

Some of the sections remain fairly static throughout the life cycle of the laser application; others will need to be updated regularly. Therefore, it is important that the file represents the current status of laser safety. It could be audited by external organizations, including regulatory officers. The impression they get of your organization could be colored by what they find in the file. Much of real-life inspections hinges on perception. Some sections of the binder are for reference and therefore for appearance. A well-kept program binder provides a good impression of laser and safety awareness and concern.

A benefit that comes with time is that the regulatory officers begin to recognize the format of the laser-safety management program files. Even the existence of the file produces a positive impression.

The following presents guidance on what to include in each section. The actual contents will depend on your development of the laser-safety management program and are likely to change with time as you find out what works best for you.

6.4.1 Standard Operation Procedures

The first section of the binder provides an opportunity to describe the laser application, hazards, and control measures. What makes a good standard operating procedure (SOP) is covered in Chapter 4. The level of detail will depend on the level of the laser-safety management program: corporate, application, or somewhere in between.

6.4.2 Training and Authorized Worker List

This section should contain a list of all training courses required by this SOP. The courses can be listed in a spreadsheet matrix. All staff working under the SOP are required to sign the matrix. This shows they have read the SOP and have completed the training requirements. The principal investigator (PI) may add or delete names and course requirements for individuals as needed. This is a simple way for the responsible individual to keep track of training and demonstrate that all staff have read the SOP.

6.4.3 Interlock Checks

The laser use area may require access controls such as door interlocks. There may also be other safety devices that require periodic inspection or verification of operation. This section of the binder is the place for documentation related to such inspections to be kept. More than just the log showing a history of dates needs to be recorded; so do any procedures that need to be followed to perform the inspection. A copy of the periodic (set within the SOP) interlock checklist is found here. Make entries on this form and continue on the same page until the form is filled (see example at end of chapter).

6.4.4 Alignment Procedures

When one considers that over 60% of laser accidents that occur in the research setting occur during the alignment process, it is easy to see why a section in this binder should be dedicated to alignment. In this section either generic alignment guidance (see example in Chapter 4) or specific alignment procedures for the laser operation should be found. While specifics are desired, the general alignment is adequate (if followed).

6.4.5 Audits and Self-Assessment Reports

The auditing of the laser use area is critical to good laser safety and is a required task of the LSO or designee. This section contains a copy of such audits. If management or users perform any self-assessments or other audits, those records should be here (this does not include regulatory inspections). Program management or self-inspection should be encouraged. Specific status or required corrective actions may be found here. Auditing is a very powerful tool to demonstrate that you are doing what you say you are doing. This section should include some kind of audit plan. You may decide to dedicate a specific time to audit the complete laser-safety management program or you may break the audit down into a number of smaller tasks that are carried out on a regular basis throughout the year.

There will be results from the audits. They may indicate that everything is still completely up to date and the operational use of the laser-safety management program correlates with what actually happens. If so, this is still a valuable result that needs to be recorded. If there are noncompliances or something needs to be improved, then this needs to be recorded and an action plan developed.

The frequency of auditing specific aspects of the laser-safety management program will depend on the risk if a given part of the program fails. For example, a safety-critical control measure may need to be audited frequently — some even each time the laser is used.

Should you be visited by an enforcement officer who is assessing your laser safety, then this part of the file can be extremely useful. If, for example, you have an interlocked enclosure and someone manages to get exposed to the laser beam because the interlock did not work, can you demonstrate that the interlock had only recently failed? If the interlock checks were part of a weekly safety audit, then you may be able to demonstrate that it was working at the last check. This will generally only convince the enforcing officer if there is a written record of the test, signed and dated. If you have not been carrying out the safety audit, then it is possible that the interlock failed shortly after installation.

6.4.6 Material Safety Data Sheets and URL

Material safety data sheets (MSDSs) are for chemicals. The most current MSDSs are accessible on the Web, and it is recommended that a Web address be listed here. For some research labs, it would be easy to fill a binder just with MSDSs. Workers may wish to keep hard copies here for particular chemicals used under this SOP.

6.4.7 Checklists

This section is available for additional checklists that maybe needed under the SOP.

6.4.8 Engineering Safety Notes

If specialized equipment such as vacuum vessels is used for the project, this section allows one to keep a copy or reference where any engineering or safety documentation exists to verify the manufacturer of the equipment.

6.4.9 Contact List

This section contains useful contacts for various EH&S concerns. It should be updated annually or the information may not be accurate, for instance, if safety personnel change. The PI may add extra names applicable to the SOP or facility. Names and contact details should be included for everyone involved in laser safety, appropriate to the level of the laser-safety management program.

6.4.10 ES&H Chapters

Most institutions have a health and safety manual. Such a manual lists the hazards and requirements for working with such. Examples would be lasers, electrical systems, toxic gas, and radioactive materials. This section contains hard copies of the most relevant chapters of the health and safety manual. One could only list Web address, but having a hard copy for staff to reference generally makes it more likely to be used.

6.4.11 Lessons Learned

This section contains information and possibly copies of lessons learned applicable to this SOP activity. Because lessons learned is a strong laser safety reinforcement tool, providing lessons learned for users can only help the laser safety program.

6.4.12 Emergency Procedures

This section contains general institution emergency response information. In addition, the PI may add SOP-specific emergency instruction and his or her own emergency response contacts here.

6.4.13 Miscellaneous

There will invariably be things that do not belong in any of the other sections. They could be filed in this section. However, as part of the audit of the file, it may be obvious that some of the items that have accumulated in this section could form a new part of the file. If one of the other sections is not being used, then you could move the information from here to there.

6.4.14 Optional Additions

6.4.14.1 Safety Structure

The safety structure for the organization may be a published document or organizational chart. Therefore, that document could be inserted in this part of the binder or there could be a reference to it. However, more detail is required in this section.

6.4.14.2 Specific Procedures

A number of procedures may be involved or related to the work in this laboratory and SOP. Procedures and instructions on how to perform them would be found in this section. In addition, a list of approved or authorized personnel to perform each procedure should be listed.

Many aspects of laser work require written procedures, written systems of work, local rules, or permits to work. In essence, safety rules and instructions may be needed. The detail will depend on the part of the life cycle that you are involved with. It is suggested that the life cycle be drawn up for the laser application and the parts of the life cycle that need written procedures to be identified. Similar procedures may be required for more than one part of the life cycle and it may be possible to combine these.

If some parts of the cycle involve external organizations, such as service companies, then your procedures may include some sort of handover documentation, both at the start and finish of the work. The actual safety procedures for the intended work, which include how the equipment is confirmed to be back in the normal working condition, should be provided by the external organization.

6.4.14.3 Safety Committee Minutes

If a Laser Safety Committee or group meetings exist, safety minutes and issues would go into this section. It is important that Laser Safety Committee minutes be taken and documentation include corrective actions, not just problems.

6.4.14.4 Standards and Regulatory items

If any regulatory activity for the operation is described in the SOP, that documentation would be found here.

6.4.14.5 Control Measures

This section of the file is used to state what control measures are in place and, just as importantly, the rationale behind the control measures. For example, in our experience, some establishments have laser safety goggles that were purchased years ago and no one can remember why a particular pair of goggles was bought. The stated optical density bears no relation to what appears to be necessary. Were the goggles the only pair available and been passed for user to user, or did the originator know something that has been forgotten in the mist of time?

6.5 INSTITUTIONAL BINDERS

In the preceding sections we discussed a laser-safety program binder and safety binder for the individual SOP. It may be to the institution's advantage to also keep an organizational binder. The LSO would be the logical person to maintain and hold such a document. That binder should contain the following sections:

1. Safety structure
2. Safety committee
3. Standards and regulations
4. Training program
5. Audits and reports

6. Risk assessments
7. Control measures
8. Incident investigation
9. External liaison
10. Miscellaneous

6.5.1 Safety Structure

The safety structure for the organization may be a published document or organizational chart. Therefore, either that document could be inserted in this part of the binder or there could be a reference to it. However, more detail is required in this section.

6.5.2 Safety Committee

In larger organizations, or organizations with a large number of lasers, it may be beneficial to have a laser safety committee. This allows laser safety staff to share experiences and keep up to date with changes to laser safety standards or regulations.

It is surprising how often laser safety is treated in complete isolation from the general safety system of an organization. This should not happen. As a minimum, the LSO, or whoever has overall responsibility for laser safety, should have a route to the general safety committee. Depending on the hazards associated with the laser application many general safety issues could be relevant. This section of the file can be used to outline the committee structures, record minutes of meetings (or have a pointer to their location), and provide a record of required actions.

6.5.3 Standards and Regulations

A growing number of standards and legal documents affect our work with lasers. This section of the file is a reference section in which to record the appropriate standards, regulations, and guidance for your particular laser application. It could be used to pull out sections of those documents that are particularly relevant to you. There may be some cross-referencing to other parts of the laser-safety management program file. This section of the file could be used to store the actual documents; it is more usual to list references to the location of the documents.

6.5.4 Training

This section should contain a list of all required training courses related to laser work, in particular, the course outline of the basic laser safety course and any medical surveillance requirements and polices.

6.5.5 Audit Procedure and Records

Auditing is a very powerful tool to demonstrate that you are doing what you say you are doing. This section should include some kind of audit plan. You may decide to dedicate a specific time to audit the complete laser-safety management program, or you may break the audit down into a number of smaller tasks that are carried out on a regular basis throughout the year.

6.5.6 Risk Assessments

The risk assessments should form the basis for laser-safety management. This section of the file should either contain the actual risk assessments or references to their locations. More importantly, there should be an action plan. This is a document that takes the conclusions from the risk assessments and, where actions are required, logs them, complete with who is responsible and when the actions need to be completed. This can be thought of as identifying the current finish line for our road to successful laser-safety management and the map of how to get there.

6.5.7 Control Measures

The risk assessment will take into account the control measures that are in place. Here the institution would list its generic control measures for an area using class 3B or class 4 lasers. The risk assessment may identify additional control measures that need to be implemented.

Interlocks, and interlock override systems, are used in many research facilities. When questioned about the reason for the choice of a particular way of working, sometimes no one can remember. The person who made the decision may have left long ago. This section of the file attempts to overcome this problem by requiring a clear statement of control measure policy and how it is implemented.

6.5.8 Incident Investigation

All laser incidents should be investigated, and that policy and protocol should be listed here. You may wish to point to where such records might be kept or you can keep them in this section. Should you be visited by a regulatory officer who is assessing your laser safety, then this part of the file can be extremely useful. If, for example, you have an interlocked enclosure and someone manages to get exposed to the laser beam because the interlock did not work, can you demonstrate that the interlock had only recently failed? If the interlock checks were part of a weekly safety audit, then you may be able to demonstrate that it was working at the last check. This will generally only convince the enforcing officer if there is a written record of the test, signed and dated. If you have not been carrying out the safety audit, then for all you know, the interlock may have failed shortly after installation. This section is also a good location for lessons learned.

6.5.8.1 Lessons Learned

Lessons learned are a valuable tool for laser safety. This section would be a copy of any lessons learned that the LSO or institution has distributed to laser users.

6.5.9 External Liaison

This section of the file is a bit like a directory of people and organizations you deal with in relation to the use of the laser. It may include details of suppliers of laser products, the manufacturer if other than the supplier, the suppliers of control measures, and perhaps details of contacts within enforcement authorities and standards bodies.

6.5.10 Miscellaneous

Invariably, some items will not fit into one of the other nine sections. These could be filed here. However, as part of the audit of the file, it may be obvious that some of the items that have accumulated in this section could form a new part of the file. If one of the other sections is not being used, then you could move the information from here to there. Alternatively, you may need to move to a 12- or even a 20-section file.

Repeating this program management binder is essential to laser safety. It acts as the core element to users and regulatory bodies.

6.6 SAMPLE FORMS

1. SOP: Chapter 4
2. Training matrix: Chapter 7
3. Interlock log: this chapter
4. Alignment guidance: Chapter 3
5. New user safety checklist: this chapter

6.6.1 Sample Safety Check List

This checklist must be completed during the first week of employment. The purpose is to ensure that the new staff member or student is aware of the proper use and location of safety items and what actions to take in case of an emergency. Once complete, place the form in the laser management binder (see Figure 6.1).

	YES	NA
1. Been instructed to read SOP (any work authorizations)	[ ]	[ ]
2. Orientation to hazards found in work area	[ ]	[ ]
3. Location of nearest fire extinguisher	[ ]	[ ]
4. Location of phone & emergency phone number	[ ]	[ ]
5. Given handling & waste disposal instructions	[ ]	[ ]
a. Radioactive	[ ]	[ ]
b. Chemical	[ ]	[ ]
c. Instructed on chemical/solvent storage	[ ]	[ ]
7. Shown building emergency assembly point	[ ]	[ ]
8. Location of eyewash stand	[ ]	[ ]
9. Explanation of what eyewear to use and when	[ ]	[ ]
a. Laser eyewear per wavelength(s) in use	[ ]	[ ]
b. Safety glasses	[ ]	[ ]
c. Face shield	[ ]	[ ]
10. Location of protective eyewear	[ ]	[ ]
11. Explanation of interlock system (if applicable)	[ ]	[ ]

Signature of employee Print Name

Signature of PI (or designee) Date

FIGURE 6.1 Sample safety checklist.

7 Laser Safety Training

Somehow in safety, training is always part of the answer.

7.1 USER TRAINING

A long-established path to laser safety is training users and those working around lasers. Effective laser safety training is not just a review of laser safety principles but a cultural change in the students — one that the instructor and institution hope they embrace. They must understand that laser safety responsibility does not start and end with themselves but extends to all those working around them. An example of this can be seen in society's acceptance of motor vehicle seat belt, a safety device that for a considerable period of time met with great user resistance. Today the use of seat belts is an automatic reflex for most drivers and passengers. This type of cultural change has still not happened among laser users. It is common for users to think that if anything happens, they alone will pay the price.

Traditional laser safety falls into three categories: users, the laser safety officer (LSO), and incidental personnel. LSO training is generally the same as or similar to the user group, but I would argue that these individuals need a greater level of safety training. User and LSO training is needed for the highest-class lasers available to the users at the facility the LSO is responsible for. Typical topics covered in laser safety training are the following:

1. Bioeffects of laser radiation on the eye and skin
2. Significance of specular and diffuse reflections
3. Laser and laser system classifications
4. Control measures
5. Overall responsibilities of management and employees
6. Fundamentals of laser operation (physical principles, construction, etc.)
7. Nonbeam hazards of lasers
8. Medical surveillance practices (if applicable)

Of these eight topics, a strong case can be made that the first five are critical to user. Training should be tailored to the type of lasers and wavelengths the user is expected to come in contact with. At an R&D facility, a comprehensive approach is well worth the time and effort.

In the classic mode, LSO training would also include:

1. Laser terminology
2. Types of lasers, wavelengths, pulse shapes, modes, and power and energy
3. Basic radiometric units and measurement devices
4. MPE levels
5. Laser hazard evaluations and other calculations

To obtain a higher professional level, an LSO's training should also include all the following:

1. Accident investigation
2. Technology updates germane to one's facility
3. Understanding of optics
4. Roles of regulatory agencies

7.2 AWARENESS TRAINING

Awareness training is usually developed for incidental staff, that is, people who may be around laser equipment but are unlikely to be exposed. That does not mean these people may not have a high perception of exposure. This perception issue should be a major concern to users and the LSO. More regulatory inspections are started by ill-informed workers' concerns that are sent to government agencies than by regularly scheduled audits.

Awareness training does not have to be a burden; rather it can be added to routine new-employee orientation, which most firms do. Such training can consist of a few slides about laser signs and warning lights, along with any restrictions in designated laser areas. If flash lamp light is a common occurrence, it should be explained to incidental staff so they do not think they are being exposed to laser radiation.

Next we will discuss specialized awareness training on expanded beams and their new classifications.

7.2.1 Class 1M, Class 2M, and Class 3R Awareness Training

This section covers optional training to users of class 1M, class 2M, and class 3R laser systems. What would be needed is a simple, brief program designed for easy implementation by persons other than LSOs or education instructors, such as first line supervisors. Potential topics could include:

1. Simple explanation of a laser
2. Comparison of laser light and ordinary light
3. Statement cautioning against intentional overcoming of the human aversion reaction and staring into a class 2 or class 2M laser beam

4. General explanation of the various lasers classifications
5. Description of the nature of near infrared (NIR) beams where applicable
6. Explanation of class 2, 2M, 1M and 3R lasers and the relative potential hazards of each
7. Explanation of the potential for collecting and focusing optics to increase the hazard

7.3 ON-THE-JOB SAFETY TRAINING

Note that this text uses the term *basic laser safety training*. We use this term because laser users should receive on-the-job, specific orientation to the laser hazards in the areas they are to be working in. Without this specific training, they are not fully prepared for laser work. Learning how to turn on the laser equipment is not part of hazard orientation. Hazard orientation should cover items such as where stray reflections might be expected, how to use beam location devices, and strategy for beam containment. One might even have a system for procedure training prior to letting new employees work alone. This all takes valuable equipment and research time. Each worker should be well trained to perform his or her job tasks.

7.4 EXPLORING TRAINING OPTIONS

Traditional basic laser safety training has been conducted in a classroom-style presentation. Technology today gives the LSO several options on how to present laser safety information. These include videotapes, computer-based training (prerecorded CDs), text-based and Web-based (also known as computer assisted) training, and lecture. Each of these have pros and cons.

7.4.1 VIDEOTAPES

While this method of training is seen less and less, it is still viable. Its benefit is flexibility of access. Drawbacks are that videos cannot be updated and in most cases are too generic and cover too much material but not enough of the users' applications. It is strongly recommended that a paper test be given along with the tape. The tape and test combination has several strong points; it provides proof of understanding, and if there is a language comprehension problem, it should be revealed. The test may force the user to observe the video more than once. One last advantage to some LSOs is that they do not have to do the presentation.

7.4.2 TEXT-BASED TRAINING

This involves giving out a booklet with pertinent laser safety information. This can be customized and updated as needed. The handout can also serve as a reference source. Once again, an examination is recommended to accompany the handout. The document should be sent to the LSO for grading and to trend areas of lack of understanding. Do not distribute a handout and only require people to sign off that they have read it; this is a recipe for ineffective training.

7.4.3 Computer-Based Training

This is very similar to video training, but sometimes it can be edited. It offers flexibility and results can be linked to a central training system. The drawbacks are that computer presentations are hard to edit, and once again are generally too generic. They do not allow the student to raise questions. Also, students' attention sometime wanes. This is why an examination is advantageous.

7.4.4 Web-Based Training

The success of Web-based courses rests with the skills and imagination of the computer support staff who create the course and the technology knowledge of the subject matter expert who supplies the information. The technology is available to make such a course interactive and engage the student. Once again, an exam is required.

7.4.5 Lecture (Classroom Style) Training

While this can be the most effective means of laser safety training, it has several pitfalls and drawbacks. The most common objection to a lecture class is time: time for the class and the cost of students' time away from their duties. Another difficulty is language barriers, but this approach, if taught by the LSO, allows the LSO and the users to meet and ask questions of each other and make a connection.

One item that is recognized by standard bodies is how trainer qualifications are ensured or checked. This can only be done by having someone from management or training audit the course and help the instructor improve his or her technique.

7.5 NEW TRAINING SUGGESTION: A LASER LESSONS LEARNED CLASS

The concept of this type of class was first presented at the International Laser Safety Conference in 2005. The object of the laser lessons learned course is not to repeat the Web course but to present laser-related lessons learned to your staff. In this author's opinion, lessons learned is the strongest safety reenforcement one can present to the laser use community. It can show how a practice that might be common to laser users can lead to a dramatic injury and a programmatic long-term work stoppage.

The course outline is as follows:

- Whom accidents happened to
- Commonly given reasons for violating laser safety procedures
- A review of perception incidents
- A review of several laser accidents
- What to do if an injury is suspected and what we can do about injuries
- Discussion of whether laser accidents can be stopped
- Summation

Such a class will receive positive feedback from experienced and inexperienced laser users. The class draws out comments from laser users on the times and reasons laser safety practices were not followed. Major laser accidents that relate to a research setting are discussed, along with who is involved in accidents. The examples used state where the accidents happen and the specification of the lasers. All this is done to clearly relate to our users the common elements between their laser use practices and those involved in the accidents.

The examples presented in the class show how other laser users have been involved in laser incidents and explain deviations from laser safety practice and the consequences, including the consequences to individuals and programs. Most importantly, the class discusses how compliance with existing protocols would have prevented the incidents. While basic and refresher laser safety training courses are important, learning how others have been injured makes a lasting impression. As our laser use changes, the class is flexible enough to use lessons learned from these new applications and directions. A class dedicated to laser lessons learned is a successful means to laser safety.

"Experience is not what happens to a man. It is what a man does with what happens to him."

Aldous Huxley

7.6 REFRESHER TRAINING

A lessons learned class could be considered a refresher class. The goal of a refresher class is not to require users retake the basic class every few years, but rather to receive a laser safety booster shot. This is an ideal application of Web-based or computer-based training. One is looking for a short presentation or reminder of laser safety issues. Opinions on frequency vary, from every 2 years for radiation training to 3 to 5 years. Which is best is best determined by someone familiar with the user population.

7.7 LASER QUIZZES

We have already mentioned the value of laser quizzes to determine comprehension of or alertness to a training mechanism. If fewer than 100% or your laser users pass your laser safety course, you should give feedback to students who either fail or get some questions wrong. The feedback can be in the form of a memo that highlights the questions they got wrong and explains the correct answer. One could start the memo by indicating there was a miscommunication or that the student did not understand the issues and supply the rationale for the correct answer.

If many laser users get the same question(s) wrong, this is a call for the LSO to review the material and see how the information can be made clearer. Below are a number of basic laser safety sample questions.

7.7.1 Sample Questions

1. Which class of laser presents the greatest hazard?
   a. Class 1
   b. Class 2
   c. Class 3b
   d. Class 4

2. Ultraviolet radiation biological effects:
   a. Exhibit a delayed effect
   b. Occur immediately
   c. Have no negative affects
   d. Only affect fair skinned people

3. Traditionally, which control measure is most effective?
   a. Administrative
   b. Procedures
   c. Engineering
   d. Protective equipment

4. Class 2 lasers must be invisible.
   a. True
   b. False
   c. Makes no difference
   d. I am color blind, cannot tell

5. ANSI stands for:
   a. American National Standards Institution
   b. All Natural Standards Institute
   c. American National Standards Institute
   d. Allstate Nation Standard Ignition

6. The laser safety officer position:
   a. Is a recommended position
   b. Must be full time

c. Must be on-site at all times lasers are used
d. Can be part time

7. A requirement for Class 3b and 4 laser users:
a. Is training
b. Is training and medical surveillance
c. Is medical surveillance
d. Is mediation

8. ANSI Z136.1 title is:
a. Safe use of lasers in medical facilities
b. Safe use of lasers outdoors
c. Safe use of lasers
d. Safe use of lasers in communications

9. Which agency is responsible for setting laser safety product codes?
a. OSHA
b. Cal-OSHA
c. CDRH
d. DOE

10. All laser safety eyewear must at a minimum labeled be with its:
a. Effective wavelength
b. Effective optical density
c. Wavelength range and optical density
d. Manufacturer's code

11. An ultrafast laser pulse will have what effect on laser eyewear?
a. None
b. Go right through
c. A portion may bleach through
d. Combust

12. Laser accidents most commonly occur during which activity?
a. Normal operation
b. Service
c. Alignment
d. Maintenance

13. Keeping spare optics on optical tables is:
a. Recommended
b. A potential source of unplanned reflections
c. An easy storage strategy
d. The way it will always be

14. Laser door interlocks provide safety to:
    a. The laser user
    b. Unauthorized users
    c. No one
    d. Wandering pets

15. The retina generally is *not exposed* to what kind of radiation:
    a. Visible
    b. Ultraviolet
    c. Near infrared
    d. a and c

16. The signal words on laser warning signs are:
    a. Caution, Notice, Danger
    b. Caution, Beware, Danger
    c. Unattended, Danger, Caution
    d. Laser on, Laser off

17. The majority of laser operator deaths are result of:
    a. Overeating
    b. Flooding
    c. Robotic error
    d. Electrocution

18. List four nonbeam hazards:
    a. ______________________________
    b. ______________________________
    c. ______________________________
    d. ______________________________

19. List three good practices during laser alignment:
    a. ____
    b. ____
    c. ____

20. "Laser Danger" warning signs are what color?
    a. Blue
    b. Yellow
    c. Green
    d. Red

21. As a rule of thumb, which has the greatest nominal hazard zone?
    a. Direct viewing
    b. Specular reflection
    c. Diffuse reflection
    d. Backward reflection

22. Laser safety training is required for users of which class of laser(s)?
    a. Class 1
    b. Class 2
    c. Class 3a
    d. Class 4

23. Might a temporary control area be needed around a Class 1 laser product?
    a. Yes
    b. No

24. Can OSHA's lock-out tag-out rules impact laser use?
    a. Yes
    b. No

25. Which is an engineering control?
    a. Laser eyewear
    b. Protective housing interlocks
    c. Warning signs
    d. Standard operating procedures

26. The retina generally is exposed to what kind of radiation?
    a. Visible wavelengths
    b. Far infrared wavelengths
    c. Near infrared wavelengths
    d. a and c

7.8 THE TRAINING MATRIX

Laser users join and leave the R&D group, and new training requirements arise. How does the principal investigator keep on top of this fluid situation? One answer is a training matrix. A matrix such as the one shown in Figure 7.1 is a visual means to determine if training is current and if the list of laser users is complete.

7.9 LASER POINTER AWARENESS

Laser pointers seem to go in and out of the headlines. Just when one thinks pointers no longer require any attention, they hit the news again. Therefore, if laser pointer awareness education is determined to be desirable, suggested topics can include:

1. Simple explanation of a laser
2. Comparison of differences between laser light and ordinary light
3. Precautions for use
4. Effects of exposures

5. Misuse
6. FDA warning on misuse of pointers
7. FDA limit of 5 mW
8. Local ordinance limitations

Illustrating the seriousness of laser pointer safety, here is a December 18, 1997, posting from the CDRH web page (http://www.fda.gov/bbs/topics/NEWS/NEW00609.html).

FDA Issues Warning on Misuse of Laser Pointers

The Food and Drug Administration is warning parents and school officials about the possibility of eye damage to children from hand-held laser pointers. These products are generally safe when used as intended by teachers and lecturers to highlight areas on a chart or screen. However, recent price reductions have led to wider marketing, and FDA is concerned about the promotion and use of these products as children's toys.

The light energy that laser pointers can aim into the eye can be more damaging than staring directly into the sun. Federal law requires a warning on the product label about this potential hazard to the eyes.

"These laser pointers are not toys. Parents should treat them with the appropriate care," said FDA Lead Deputy Commissioner Michael A. Friedman, M.D. "They are useful tools for adults that should be used by children only with adequate supervision." FDA's warning is prompted by two anecdotal reports it has received of eye injury from laser pointers — one from a parent, the other from an ophthalmologist. Momentary exposure from a laser pointer, such as might occur from an inadvertent sweep of the light across a person's eyes, causes only temporary flash blindness. However, even this can be dangerous if the exposed person is engaged in a vision-critical activity such as driving.

Green pointers have added to the pointer safety confusion that may come up in training sessions. Once again from the FDA:

> FDA is concerned about recent reports of laser products directed at aircraft — a potentially hazardous situation. The agency is particularly concerned about the increased availability of overpowered green laser pointers. Overpowered green laser pointers are those that may have been modified to emit more radiation than the manufacturer's original product.

Training Matrix

SOP # or Lab Name

Compliance code
NC Required, not compliant
Date entered when taken
NA Not required by person
Check for course expiration

This matrix list all authorized user of lasers in this lab.
In addition all required training courses are listed
It is role of PI to confirm training and see staff has taken required training. Some staff can be excluded from some courses based on their work activities

Sample course listing

Date form generated	Annual review user list		Eye exam		Waste	H. H. Com.	Pressure Safety	Lasers	Capacitors	Electrical	High Voltage	Lock & Tag	Energized Equip.	Back Care Workshop	
		Frequency	0		1	0	5	5	5	5	5	5	5	0	0
Name	Signature	Date			-RW	WEB		WEB	WEB	WEB	WEB	Lecture	Lecture		WEB

FIGURE 7.1 Training compliance chart (training matrix).

8 Personnel Protection Equipment*

8.1 INTRODUCTION

The most common misconception laser users have about laser protective eyewear is that it is the first line of defense against laser radiation. In reality it should be the last line of defense. Beam containment will do more for a laser user than laser protective eyewear. If you can eliminate the possibility of eye damage by enclosing the laser beam path such that *no* radiation exposure to the eye is possible, then do so. While critically important, the implementation of laser protective eyewear is always understood to be the second line of defense. Laser protective eyewear has a valuable role to play in laser safety and presents many challenges to the user and laser safety officer (LSO). This chapter covers the selection and use of laser protective eyewear (see Chapter 5 for additional discussion).

Laser protective eyewear comes in two types: full attenuation and alignment eyewear. Full attenuation means that no visibility of the termination point of the beam (or inadvertent intrabeam exposure) is feasible when wearing the laser eyewear. Conversely, alignment or partial attenuation laser eyewear allows an individual to see some of the termination point of the beam for various purposes such as beam collimation, laser beam path alignment, and so on.

Frequently, one encounters cases where LSOs recommend and researchers are then supplied with full attenuation laser eyewear that subsequently is underutilized because of research conditions where partial attenuation is required for proper execution of laser-related applications.

8.2 FULL ATTENUATION

Without exception, for class 4 lasers and class 3b lasers (when the maximum permissible exposure [MPE] limit is exceeded), it is recommended to provide full attenuation laser protective eyewear in all ultraviolet (UV) (i.e., nominal 190 to 380 nm) and ocular focus near infrared (NIR) nonvisible (i.e., nominal 700 to 1400 nm) wavelengths, as well as mid to far IR regions. The logic in doing this is that if one cannot see the beams and they exceed the MPE limits, then there is no reason to do anything other than fully attenuate those same wavelength regions.

Moreover, in the visible regime (i.e., nominal 400 to 700 nm) when detection of the termination point of the visible laser wavelength is *not* required for one's application, then full attenuation of these same visible wavelengths is also recommended.

* Major contributions to this chapter come from Will Arthur and Cathi Scogin of Glendale eyewear.

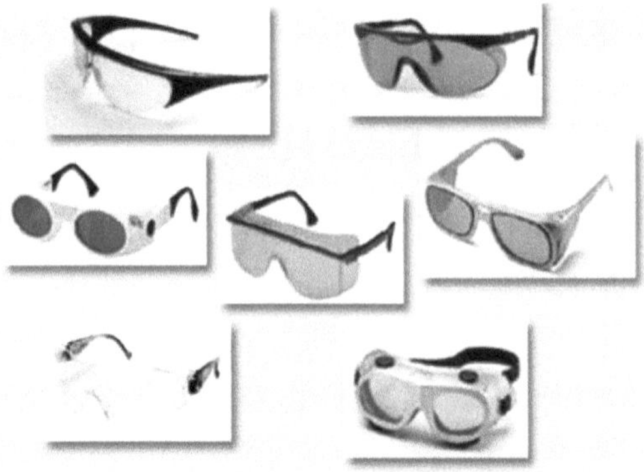

FIGURE 8.1 (See color insert following page 164.) Laser eyewear, Glendale laser products.

The LSO has the task of recommending proper eyewear selection for the wavelength or wavelength region in question to meet the required optical density (OD) for each laser application (Figure 8.1). Once the small source intrabeam OD for each laser wavelength or wavelength region has been calculated, various other ancillary conditions emerge that may positively (or negatively) impact the intended use of the chosen laser protective eyewear.

To be effective, laser eyewear *must* be worn. As obvious as this statement is, the single most prevalent cause — by far — of laser-related eye injuries is that laser protective eyewear, while is typically available and appropriate to the prevailing laser application, was not worn.

Why? This is where many of the ancillary features such as weight considerations between glass and polycarbonate lenses, acceptable versus unacceptable visual light transmission (VLT), subjective preferences of comfort and fit, prescription lens (Rx) capabilities, propensity of eyewear to fogging, and peripheral visual capacity or lack thereof come into play.

8.3 VISUAL LIGHT TRANSMISSION

Undoubtedly, VLT and fit are the two most compelling factors in the usage or aversion to usage of laser eyewear. Simply stated, VLT is the mean average percentage of the entire visible spectrum, as weighted for blue spectral responsiveness, that is *not* filtered by these lenses. Repeatedly, experience has indicated that the higher the VLT, the higher the likelihood of eyewear usage and consequently laser eyewear safety compliance.

In many research and academic circumstances, overhead room lights may be turned off for a variety (e.g., beam collimation, alignment, etc.) of conditions, and VLT in these circumstances becomes a pre-eminent concern. Moreover, laser-related electrical hazards, which have caused serious injuries possibly including death, must be fully considered in light of diminished visual acuity resulting from a loss of VLT when wearing laser protective eyewear. Lest we forget, while laser radiation can blind you, electricity can kill you.

Additionally, the distinction between OD and VLT, especially in full- attenuation conditions, is sometimes misunderstood or misrepresented. Assumptions abound that a higher OD *necessarily* implies a reduction of VLT. However, reduction of VLT is directly correlated to a higher OD only when visually limiting optical densities are directly attributable to the visible (i.e., nominal 400 to 700 nm) region only.

In laser eyewear attenuation conditions in UV; near, mid, and far IR regions; or a multiwavelength combination thereof, one may encounter eyewear that posseses high OD, low VLT; high OD, high VLT; low OD, low VLT; or low OD, high VLT. In my estimation, any eyewear possessing a VLT at less than fifteen to twenty percent (15 to 20%) is dangerously close to creating a loss of visual acuity and causing other potential (notably electrical) dangers to become considerably more likely.

Therefore, in seeking full attenuation laser eyewear with appropriate OD values for one's application, increasing VLT may require certain trade-offs. Typically, this is the decision juncture at which one considers the use of plastic versus glass lenses. Polycarbonate lenses are lighter than glass lenses. As such, polycarbonate lenses have inherent (and perfectly logical) preferability to the user, especially in conditions where protracted usage is required. There are certain common and very prevalent wavelength regions (notably Nd:YAG at 1064 nm) in which glass lenses have higher VLT than polycarbonate lenses. In this instance, the trade-off is that while one is increasing the VLT, one is simultaneously increasing the weight of the eyewear and thereby potentially diminishing its perceived comfort.

Fortunately, various manufacturers of both glass and polycarbonate eyewear have noted the general preference for polycarbonate eyewear and have made significant strides in increasing their products' VLT in Near Infrared Radiation (NIR) and certain other visible wavelength regions.

8.4 COMFORT AND FIT

Comfort and fit considerations are the wholly subjective and depend entirely upon individual preferences that each wearer maintains concerning how a specific set of laser protective eyewear feels when worn. Comfort and fit primarily center upon personal preferences issues, such as overall comfort when evaluated in terms of short, moderate, or protracted wearing times. If a pair of protective eyewear does not fit properly, not only can it not perform its function to the required specifications, but the likelihood of it being used decreases. This is true for a respirator, facemask, and laser protective eyewear.

Users want their eyewear to be as natural an extension of their faces as possible. They do not want to be constantly reminded that they are wearing eyewear by its being too loose, too tight, or too heavy, fogging up, slipping, or other common problems. Therefore, finding properly fitting eyewear is well worth the time. One size does not fit all. One solution may be to place a strap across the back to keep the frame as tight as necessary. Another solution may be flip-downs on one's own glasses. Manufacturers offer a range of options in sizes, including new eyewear for slim faces and for very large faces. There are options for fitting different nasal profiles, including flat or low nasal profiles, and combinations for small faces with flat nasal profiles. Adjustable temple lengths are also helpful, as well as temples with gripping ends. Bayonet temples (the straighter temple) also help in fitting large faces. Choices of laser protective eyewear have come a long way. All users should be able to find just that right pair.

8.5 DAMAGE THRESHOLD CONSIDERATIONS

Once you find the appropriate eyewear with adequate OD to achieve full attenuation and suitable VLT, there is yet another trade-off to ponder, namely damage threshold considerations. As a general rule of thumb, polycarbonate eyewear can withstand approximately 100 W per cm^2 of direct incident laser radiation for approximately 10 seconds prior to damaging effects being noted on the lenses. Glass eyewear can withstand approximately 10 times (~1000 W per cm^2) the value of polycarbonate laser eyewear for the same time duration.

With the assumption that a collimated, focused beam is impinging upon a discrete, nonwavering point on the polycarbonate or glass lens, polycarbonate lenses are prone to exhibiting sequentially a superheated plasma effect at the surface of the lens, degradation of the absorptive dyes (with possible carbonization and darkening effects noted), the emission of smoke, possible noxious odors, the emission of flame, and potential ultimate penetration of the lenses. Glass lenses are prone to catastrophic degradation effects where the accumulation of irradiant energy results in loss of integrity, with effects such as a popping sound when the beam strikes the glass lens, potential "spider vein" crazing and, with sufficient accumulation of energy, a complete shattering of the glass lens.

Generally speaking, these physical effects for both polycarbonate and glass lenses have readily apparent visual and auditory correlates that forewarn the wearer of an impending damage threshold danger. They come into consideration when one is deciding upon which trade-offs to implement in order to optimize the likelihood of eyewear suitability and will also be discussed when ultrafast pulse considerations are presented later in the chapter.

8.5.1 Side Shields

The ANSI Z136.1 standard in "Factors in Selecting Appropriate Eyewear" mandates users to "consider" sideshields. Overall, the presence of side shields is not an issue that can be considered and then decided against. Rather, even though they may impair peripheral vision, the presence of side shields is should be mandatory and commensurate with the level of optical density that the main viewing lenses provide.

The ANSI Z136.1 standard "Safe Use of Lasers" does not require laser protective eyewear to be ANSI Z87 compliant. ANSI Z87 is the standard for safety eyewear; the most common element is impact resistance. Therefore, in evaluating eyewear, the question of impact resistance needs to be addressed.

Is it needed or not? If it is not needed, no further action is required; if the LSO hazard evaluation determines that it is required, one has three choices:

1. Obtain a pair of laser eyewear that is compliant with Z87 (most polymer eyewear is compliant).
2. Wear safety glasses over the laser eyewear.
3. Have glass laser eyewear hardened to meet Z87.

Choice 2 can affect comfort or the ease of wearing the protective eyewear and general vision, while choice 3 affects the cost of the eyewear.

8.6 PRESCRIPTIONS

Eyewear for prescription wearers has several options. These include eyewear with prescriptions ground into the glass laser lens, eyewear that holds prescription inserts, and eyewear with flips, with polymer prescriptions in the base or the flip. For ground lenses, the frame selections have been expanded to include titanium frames and frames with adjustable temples.

8.7 SENSOR CARD USE

Most common sensor cards are used to locate UV or infrared beams. The laser radiation of concern strikes the card and produces a photodynamic reaction that yields fluorescence in the visible spectrum. The users need to make sure the eyewear selected will allow visibility of that fluorescence or glow. Making the user aware of application uses and the extent of the filtration of the eyewear is extremely important.

8.8 WEIGHT

Weight of eyewear is a particular concern when acquiring multiwavelength or prescription eyewear. Depending on wavelength combination, 7-mm-thick glass is not unheard of. This thickness of glass two to three times normal prescription eyewear may prove too uncomfortable for a user to wear for extended periods. This can lead to a lack of productivity or not wearing eye protection. Breakthroughs in polycarbonate prescription flips and over-glasses may help improve this problem.

8.9 LABELING

ANSI Z136.1 and the International Electrotechnical Commission (IEC) require laser protective eyewear to be labeled with the wavelength and OD it is intended for. The laser eyewear manufacturer imprints on the eyewear the most common range of wavelengths and OD for a particular pair (Figure 8.2). For the vast number of laser users, this is satisfactory. A small segment of users use eyewear

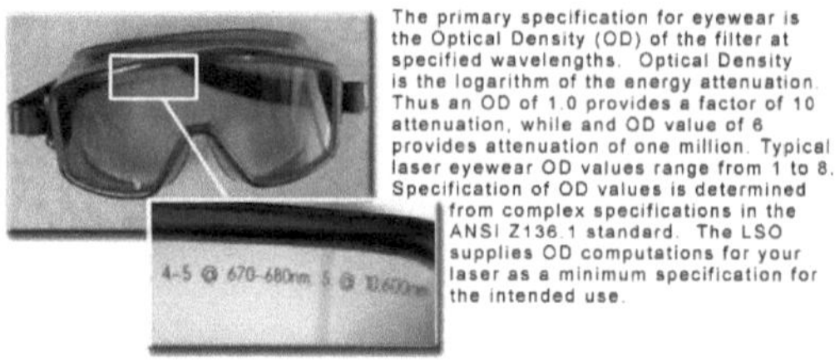

FIGURE 8.2 (See color insert following page 164.) Labeling.

for wavelengths not listed on them. Curves and other documentation provided by the eyewear manufacturer or distributor show the OD at the desired wavelength. To be compliant, the facility LSO has to label the eyewear or post the information where the eyewear is stored and have a way to identify which pair is which.

8.10 ULTRAFAST LASERS

Testing at Brooks Air Force base has shown a nonuniform bleaching effect on standard laser eyewear against ultrafast pulses. This relates back to the relaxation time of the absorption molecules. Not all eyewear for ultrafast pulses demonstrates this effect, but a significant number makes it a real concern. Therefore, ultrafast laser users who want full protection need to check the manufacturer's testing results to verify suitability of the eyewear for their use. Usually, the manufacturer can provide a sample piece of the lens for testing with a power meter in the actual application to verify the appropriateness of the lens in question.

It is imperative to recognize that with ultrafast lasers (particularly regeneratively amplified sources) there exists the potential for OD values to be compromised should femtosecond beam exposure to laser eyewear occur. Should temporary or permanent loss of OD (and commensurate exposure levels in excess of applicable MPE values) occur as a consequence of these conditions, obvious detrimental eye safety effects become possible. The core safety issue surrounding laser protective eyewear and femtosecond lasers is as follows: In certain ultrafast (femtosecond) operating conditions, saturable absorption effects with calculable losses in purported OD values of the femtosecond-subjected laser eyewear have been observed.

It is the intention of ANSI committees involved in this matter that the underlying mechanisms of the degradation effects be investigated and, to the greatest extent possible, elucidated for everyone's general understanding.

8.11 ADDITIONAL CONSIDERATIONS

Another important consideration is antifog capabilities, especially for goggles. Multiwavelength operations bring out special questions, because the more wavelengths you try to remove with one pair of eyewear, the darker the eyewear typically gets. You can try flip options or more than one pair to alleviate this problem. Laser-inscribed markings (printed ones wash off when cleaned) also help increase the longevity of the eyewear, and UV inhibitors prevent darkening over time in polymer eyewear. Finally, cost is important, but you also must consider the cost of an eye.

8.12 PARTIAL ATTENUATION (ALIGNMENT EYEWEAR)

In laser-related research applications, investigators frequently need to view the termination point of visible laser sources. These beam alignment conditions are also acknowledged to produce a notable number of laser-related eye injuries.

The purpose of alignment eyewear is to allow the user to see the beam while lowering the intensity of any beam that is transmitted through the user's eyewear to a class 2 level. To address this issue, a European norm recommends OD for alignment eyewear versus the output of lasers used.

Scale #	OD	Max Instantaneous Power/Continuous Wave laser (W)	Maximum Energy for Pulsed Lasers (J)
R1	1–2	0.01	2×10^{-6}
R2	2–3	0.1	2×10^{-5}
R3	3–4	1.0	2×10^{-4}
R4	4–5	10	2×10^{-3}
R5	5–6	100	2×10^{-2}

Therefore, for alignment laser eyewear to be effectively utilized, preferentially all of the following conditions should be in place: administrative liability acknowledgment and acceptance of it, acknowledgment of potential hazards with the utilization of eyewear that does not protect against small source intrabeam or specularly reflected exposures, and collaborative agreement between the LSO and researcher of alignment eyewear safety protocols and appropriate alignment laser protective eyewear. Once these preliminary philosophical protocols are established, the implementation of alignment eyewear can proceed.

8.13 ADDITIONAL PPE

8.13.1 Protective Windows

Laser protective windows can be thought of as large pairs of laser protective eyewear. They should meet the same requirements as eyewear regarding OD labeling. These windows are made of acrylic filter material.

8.13.2 Laser Curtains

Overall laser curtains and barriers are used as passive guards to enclose an area where class 3B or class 4 lasers are in use to protect against accidental exposure to the laser beam or for long-term blocking of laser radiation at lower power densities. Similarly, some settings using class 1 lasers, which are class 4 under service conditions, require a temporary control area. This temporary control area barrier is meant to provide a safety barrier while work continues during service. A wide variety of options exist concerning the level of protection curtains can provide. Examples follow:

Irradiated Area	Power	Density Protection Time
1 mm^2	3 MW/m^2	100 sec
500 mm^2	3 MW/m^2	20 sec
500 mm^2	1 MW/m^2	60 sec
500 mm^2	0.7 MW/m^2	100 sec

If one is up for self-testing or -certification of curtain material, many standard varieties of welding curtains provide adequate laser barrier protection. The cost difference between these materials and manufacturer-certified laser curtains could be a factor of 10 or more. One may first wish to consult with the organization's legal department prior to trying this economic approach.

9 Laser Accidents

Let's try learning from (other people's) experiences

9.1 INTRODUCTION

The use of laser technology has been increasing ever since its introduction in the 1960s. This text concerns itself with laser safety, so let us ask the questions: Are laser accidents happening? Do we have bodies to count? The simple answers are yes, laser accidents are occurring, and the number is uncertain. One survey of laser accident databases showed more than 1500 accidents from 1960 to 2002. Approximately 70% involved eye injury. A small number of reported accidents (approximately 3%) resulted in deaths. These deaths were due to either electrocution or the effects of operator room fires caused by laser radiation. Statistically this seems like a small number. As in any reporting system, laser safety professionals believe the number of laser accidents is under-reported. Also, trauma to the eye is dramatic, even if the effect passes with time. This chapter will present examples of real laser accidents. Some are equipment-related and others are the result of user error.

A number of isolated databases track laser accidents. In the United States, the Food and Drug Administration tracks equipment-related laser accidents. Other federal agencies that track their own accidents are the Department of Energy, the Army, the Navy, the Air Force, and the Federal Aviation Administration. In addition, some private industry groups have databases.

The most common factors contributing to laser accidents are:

1. Unanticipated eye exposure during alignment
2. Available laser eye protection not being used
3. Equipment malfunction
4. Improper high voltage handling methods
5. No protection provided for ancillary hazards
6. Incorrect eyewear selection
7. Inhalation of laser-generated air contaminants
8. Viewing of laser-generated plasmas (blue light hazard)
9. Improper use of equipment

9.2 ALIGNMENT ACTIVITIES

Once the laser beam path is set and established, unless you place yourself or a reflective object in the path or approach the beam with a bend magnetic around your neck, the beam will stay on its appointed path. So it can easily be seen that

FIGURE 9.1 Peer pressure can lead to accidents (author's favorite laser cartoon; see Dr. Fun Web page for others June 26, 1997 and July 2, 1997).

the overwhelming number of laser accidents or injuries occur when laser optics are being manipulated. Reflections off tools, optical mounts, lenses, or mirrors are the chief cause of random or unexpected laser beams. (See the section on guidelines to practice during laser alignment.) In addition, changes in temperature in the laser area, cooling water, and other factors can induce beam errors. The time and effort spent performing laser alignment can range from a few minutes to more than 5 hours. Temperature changes in the room, thermal expansion of optics and mounts, and dust all play a critical role in how smoothly an alignment of a laser-based system goes and how long it can take. Whether manipulations are manual or done by computer controls is another factor. A beam may disappear from the expected path, and then one has to find it and bring it back into the path. Remember that laser alignment does not relate to visible beams only, but more commonly to invisible beams, where a variety of alignment aids is needed.

The techniques for laser alignment listed below are to be used to help prevent accidents during alignment of a laser or laser system.

9.2.1 Preparation for Alignment

1. To reduce accidental reflections, remove watches, rings, dangling badges, necklaces, and reflective jewelry before any alignment activities begin.
2. Use of nonreflective tools should be considered.
3. Access to the room or area is limited to authorized personnel only.
4. Consider having at least one other person present to help with the alignment.

5. All equipment and materials needed are present prior to beginning the alignment.
6. All unnecessary equipment, tools, and combustible materials (if the risk of fire exists) have been removed to minimize the possibility of stray reflections and nonbeam accidents.
7. Persons conducting the alignment have been authorized by the responsible individual.
8. A notice sign is posted at entrances when temporary laser control areas are set up or unusual conditions warrant that additional hazard information be available to personnel wishing to enter the area.

9.2.2 Alignment Considerations

1. Whoever moves or places an optical component on an optical table (or in a beam path) is responsible for identifying and terminating every stray beam coming from that component (meaning reflections, either diffuse or secular).
2. All laser users must receive an orientation to the laser use area by an authorized laser user of that area.
3. There must be no intentional intrabeam viewing with the eyes.
4. Coaxial low-power lasers should be used when practical for alignment of the primary beam.
5. Reduce beam power with ND filters, beam splitters, or dumps or by reducing power at the power supply. Whenever practical, avoid the use of high-power settings during alignment.
6. Laser protective eyewear must be worn at all times during the activity.
7. Beam blocks must be secured (labeled if possible).
8. Have beam paths at a safe height, generally below eye level when standing or sitting. If necessary, place a step platform around the optical table.
9. The laser safety officer (LSO) has authorized eyewear with reduced optical density (OD) to allow the beam spot to be seen. Measures must be taken and documented to ensure that no stray hazardous specular reflections are present before the lower-OD eyewear is worn. Full protection OD eyewear, as listed in the laser table, is to be worn again once alignment is complete. The reduced-OD eyewear is labeled as alignment eyewear and is stored in a different location than the standard laser eyewear for this operation.
10. Skin protection should be worn on the face, hands, and arms when aligning at ultraviolet (UV) wavelengths.
11. The beam is enclosed as much as is practical. The shutter is closed as much as is practical during course adjustments.
12. Optics and optics mounts are secured to the table as much as is practical. Beam stops are secured to the table or optics mounts.
13. Areas where the beam leaves the horizontal plane must be labeled.

14. Any stray or unused beams are terminated.
15. Invisible beams are viewed with infrared (IR)/UV cards, business cards, card stock, craft paper, viewers, 3 × 5 cards, thermal fax paper, Polaroid film, or by a similar technique. Operators are aware that such materials may produce specular reflections or may smoke or burn.
16. Pulsed lasers are aligned by firing single pulses when practical.
17. Intrabeam viewing is not allowed unless specifically evaluated and approved by the LSO. Intrabeam viewing is to be avoided by using cameras or fluorescent devices.

9.2.3 Alignment Conclusion

Normal laser hazard controls must be restored when the alignment is completed. Controls include replacing all enclosures, covers, beam blocks, and barriers and checking affected interlocks for proper operation.

9.3 FAILURE TO WEAR PROTECTIVE EYEWEAR

Probably 99.99% of laser alignment accidents occur with the injured party not wearing laser protective eyewear. The excuses are common. “I cannot find a pair that fits.” “I cannot see the beam.” “They are too heavy.” “They cost too much.” “I know the set-up well enough. I don’t need them.” “I am using visible beams. I will see where the beam goes.” Many of these have a similar tone to the excuses first heard when seat belts became mandatory.

9.3.1 Solution

The solution lies in two areas: selecting the correct eyewear and peer pressure. If the eyewear selected has a comfortable fit and provides good visual light transmission and protection, many reasons for not wearing eyewear are taken out of the equation. As for peer pressure, if it is acceptable to not wear eyewear, the practice will continue. Supervisors and the laser user community in an institution have to police each other.

9.4 ELECTRICAL HAZARDS OVERLOOKED

Laser systems and electricity go hand in hand; electric shock is the second most commonly reported incident associated with laser use. Power supplies, flash lamps, batteries, capacitors, poor grounding, and accessible wires all can be found in the majority of laser systems. All need to be addressed and respected. The laser user needs to remember that between 6 and 20 mAmps is sufficient current to bring about the “let-go effect,” meaning you become part of the circuit and you cannot let go. Equipment circuit breakers are there to protect the equipment, not the users.

9.5 ADDITIONAL HAZARDS AND SOLUTIONS

A long list of additional hazards can be associated with laser work, such as pressure vessels, robotics, and so on. These will be explained and expanded on in Chapter 13.

9.6 IMPROPER RESTORATION OF LASERS AFTER SERVICING

While good work practice calls for any interlock bypass to be built so that the device cannot be restored back to service with the bypass in place, this is not the common practice. Items as thin as a paper clip or tape have been used to bypass interlocks, such as micro switches.

The solution involves a checklist, for both the service person and the user accepting the finished work. Both need to see that all safety systems are operational (including interlocks and warning lights) and that all enclosures and housings are secure.

9.7 LACK OF PLANNING

Planning means not only having all the proper tools available for the work, but having the correct eyewear and setting up warning signs and an exclusion zone to protect others in the work area.

9.7.1 Solution

Reading the procedures and developing checklists ahead of time are the best cures for a lack of planning. Many times a lack of planning goes along with rushing, which has led to numerous accidents. Checklists should list not only the activities to be performed but what tools are required to start the task. The weakness of checklists is the person or team using them. If an activity is routine, sometimes even when reading the checklist, people do not really pay attention to the list; they are anticipating the response. Every pilot knows not to take off without going through a preflight checklist. At times this becomes so rote that expected responses are checked rather than the actual instrument readings. An excellent addition to a checklist is a digital photo of each step.

9.7.2 Checklist for Developing a Checklist

1. Define the task:
   - Define the checklists and the intended user and use.
   - Know the skill and training levels of users.
   - Study relevant literature and techniques if available.
   - Discuss the checklist with experienced personnel or experts.
   - Clarify the function and task to be met by the checklist.

2. Checkpoint list:
   List descriptors for well-established criteria.
   Briefly define each of the checkpoints.
   Add descriptors for each checkpoint which requires a definition.
   Provide definitions for each of the added descriptors.
3. Sort the checkpoints; create a road map to be followed:
   Place each descriptor on a separate card.
   Sort cards by categories.
   Identify the main categories and label them.
4. Flesh out categories:
   Define each category.
   Write a rationale for each.
   Define the steps of each category.
   Present relevant warnings in case of error that apply to each checkpoint.
   Present any safety actions to be performed, for example, put eyewear on or position beam block.
   Review checkpoints.
5. Determine the order of the categories:
   Decide on order.
   Write a rationale for each.
   Review the order.
6. Perform a initial review of the checklist:
   Prepare a review version.
   Critique the checklist with potential users.
   Listen to and understand concerns and suggestions or ask questions for clarification.
   List issues in need of attention or further development.
7. Revise the checklist:
   Review concerns.
   Rewrite the checklist.
8. Format the checklist:
   Determine with users the preferred format of the checklist, that is, its appearance.
9. Evaluate the checklist:
   Perform a dry run with the checklist or with the alignment laser.
   Assess whether the checklist accomplishes its task while meeting the users' goal.

(An example: The task may be accomplished, but time to completion is excessive.)

10. Finalize the checklist:
    Consider and address the reviews of the field test (dry run).
    Print the final version.

11. Apply the checklist:
    Apply the checklist to its intended use.
    Invite users to provide feedback to the checklist developer.
12. Review and update the checklist:
    Have a mechanism in place to update the checklist as conditions change.

An excellent addition to a checklist is a digital photo of each step.

9.8 WEARING THE WRONG EYEWEAR AND IMPROPER FIT

There is not a universal pair of laser protective eyewear, at least not until the virtual imaging helmet becomes available. Therefore, users must pay attention to the specifications of the pair they are using. Is it designed for the wavelengths they may be exposed to? Is it at the appropriate OD to provide full or, in the case of alignment eyewear, partial protection? Is the eyewear in good physical shape? Have its protective features been compromised (scratches, bleaching)? Does it fit the user's face? Wearing the wrong laser protective eyewear has injured people (approximately 4% of reported accidents).

The solution is for laser protective eyewear to be labeled or coded with its OD and the wavelength range it is designed to protect you from. In some cases labeling will indicate if the eyewear is full protection or alignment, continuous wave or pulse.

9.8.1 Eyewear Failure

If one is dealing with laser beams that can burn through eyewear, why, then, are engineering controls not in place to prevent one from being in a situation where direct facial exposure is possible? Sometimes the correct eyewear was not chosen; either the OD was too low or the eyewear needed to be safety hardened but was not.

The solution is to inspect eyewear before each use. Look for scratches and signs of physical stress. Read the labeling and check for wavelength compatibility and correct OD. Evaluate the hazard from the beam; determine if remote viewing is the proper safety solution.

9.9 FATIGUE AND STRESS

These two factors can make the most diligent laser user a menace to himself or herself and others.

9.9.1 Fatigue

> My mind clicks on and off. I try letting one eyelid close at a time while I prop the other with my will. But the effect is too much, sleep is winning, my whole body argues dully that nothing, nothing life can attain is quite so desirable as sleep. My mind is losing resolution and control. (Charles Lindbergh about his 1927 transatlantic flight)

While people associate fatigue with feeling tired, few realize how serious fatigue at the workplace can be, particularly when working around photons moving at the speed of light. With very few exceptions, the average adult needs between 7 and 8 hours of sleep a night to be adequately alert for the other 16 to 17 hours that include commuting, work, and leisure and family time. Without the proper amount of sleep, the body cannot and will not function to its potential.

Symptoms of fatigue include a decreased ability to concentrate on multiple tasks, reduced ability to analyze new information, fixation, short-term memory loss, impaired judgment, impaired decision-making ability, reduced ability to think logically, daytime drowsiness and micro-sleeps, reduced motor skills and coordination, reduced ability to think critically, distractibility, reduced visual perception, loss of initiative, personality changes, depression and a feeling of indifference to one's performance, and increased reaction time.

The majority of attention to worker fatigue has been directed to two large groups: those in the transportation industry (truck drivers, pilots, and long-trip private drivers) and second- and third-shift workers. A study published in the journal *Sleep* estimated that 52.5% of all work-related accidents may be related to sleepiness. After all the investigations were complete, these disasters were attributed to fatigue: the Exxon *Valdez* oil spill, the Chernobyl power plant explosion, the *Challenger* explosion, Three-Mile Island, and Bhopal. With the rapid increase in technological capability, it is not our machines that fail us; it is the operator. Human errors are the primary cause of 90% of all accidents.

One night of disrupted sleep probably will not result in huge catastrophes, but most people do not have just one night of disrupted sleep. Night after night they try to get by on less sleep than their bodies need. Sleepiness builds into a sleep debt because the effects of inadequate sleep are cumulative. For example, assume an adult needs 8 hours of sleep each night but only gets 7. By the end of a week there is a 7-hour sleep debt, which is the equivalent of going one full 24-hour period without the proper amount of sleep. In college this is called pulling an an all-nighter. Now let us figure the sleep debt for an individual who only gets 6 hours of sleep each night (which seems to be more accurate for most Americans). At the end of the week, that sleep debt is 14 hours — or two all-nighters. The bottom line is that as fatigue increases, your risk of causing an accident increases.

The solution is simple, but in many work situations it seems as far away as the end of the day. Take breaks and get proper rest. Remember that whatever time one thinks is too valuable to use to break away and take a short rest is nothing compared to time lost to an accident. Documented cases show that operations shut down from 1 day to 4 months during accident investigations and while obtaining authorization to operate again.

9.9.2 Stress

Stress to get the job done and perform for others and pressures from management can lead to lapses in good judgment, for example, bypassing standard control measures to take short cuts. Taking these short cuts and bypassing standard operating procedures can cause serious injuries to those who know the right way to do things. It is a derivation of the saying, "Do as I say, not as I do."

9.9.2.1 What is Stress?

Webster's defines stress as a physical, chemical, or emotional factor that causes bodily or mental tension and that may be a factor in disease causation. Physical and chemical factors that can cause stress include trauma, infections, toxins, illnesses, and injuries of any sort. Emotional causes of stress and tension are numerous and varied. While many people associate the term *stress* with psychological stress, scientists and physicians use this term to denote any force that impairs the stability and balance of bodily functions.

A mild degree of stress and tension can sometimes be beneficial. Feeling mildly stressed when carrying out a project or assignment often compels us to do a good job and to work energetically. Likewise, exercising can produce a temporary stress on some body functions, but its health benefits are indisputable. It is only when stress is overwhelming, or poorly managed, that its negative effects appear.

Job stress can be defined as the harmful physical and emotional responses that occur when the requirements of the job do not match the capabilities, resources, or needs of the worker. Job stress can lead to poor health and even injury. The concept of job stress is often confused with challenge, but it is not the same. Challenge energizes us psychologically and physically, and it motivates us to learn new skills and master our jobs. When a challenge is met, we feel relaxed and satisfied. Thus, challenge is an important ingredient for healthy and productive work.

Some employers assume that stressful working conditions are a necessary evil — that companies must turn up the pressure on workers and set aside health concerns to remain productive and profitable in today's economy. Research findings challenge this belief. Studies show that stressful working conditions are associated with increased absenteeism, tardiness, and intentions by workers to quit their jobs — all of which have a negative effect on the bottom line. Recent studies of so-called healthy organizations suggest that policies benefiting worker health also benefit the bottom line. A healthy organization is defined as one that has low rates of illness, injury, and disability in its workforce and is also competitive in the marketplace.

Stress solutions can range from resetting goals to a realistic level to re-evaluating staffing levels to breaking tasks into smaller, more manageable sections. Sometimes it is as simple as the realization that just like fatigue, if stress is not recognized and dealt with, serious accidents can occur, in addition to loss of productivity and staff illness.

9.10 WORKPLACE CULTURE

A corporate or workplace culture is — like any other culture — a set of behaviors and codes that people use to govern their interactions with each other. This includes both formal, written company policies and informal rules of the road that you learn with experience. A workplace culture that calls for 12- to 16-hour work days and applies constant pressure to produce will have a history of accidents and employee health problems. In academic and medical settings the culture is well established that allows and accepts over-working and under-valuing graduate students and residents. Is it any wonder that this population (graduate students) makes up such a large portion of those involved in laser accidents and near misses?

9.10.1 Solution

Changes in workplace culture must come from the top. Risk management needs to show management how much time is lost in sick days and what the actual productivity is with such a culture. The quality of work performed and creativity is the goal — not the number of hours worked.

9.11 TO WHOM ARE THESE ACCIDENTS HAPPENING?

The majority of laser accidents do not happen to novice laser users, but to experienced laser-use personnel. Usually when those with little laser experience are exposed, they are similar to the drunk driver victim: the innocent by-stander. Approximately 70% of reported accidents are made up from the following four groups:

1. Scientists
2. Students
3. Technicians
4. Patients (number 1 group for reported deaths)

Other population categories make up the remaining 30% of the reported incidents:

1. Industrial workers
2. Doctors and nurses
3. Pilots and military personnel
4. Spectators
5. Laser show operators
6. Field service staff
7. Office staff

9.12 THE PERCEIVED INCIDENT

Sometimes an investigation will result from the perception of an injury. Even if you believe that no exposure occurred, the investigation has to be handled in a professional manner. The perception incident might demonstrate a flaw in the safety system or at the least, in training. If the investigation is handled in a manner that makes the reporting party feel as if his or her concerns are not being considered seriously, future incidents will not be reported. Worse, the next incident may be reported to a higher level or regulatory agency, all of which may damage the effectiveness of your program.

9.13 ACCIDENT INVESTIGATION

For the LSO or laser safety advisor, the goal is to prevent laser accidents and incidents, but one must be prepared to investigate an accident or incident that is reported. Below are the elements of accident investigation.

9.13.1 Who Should Lead the Accident Investigation?

The LSO should be the lead contact on a laser-related incident. Many LSOs also perform several other functions. Therefore, it might be necessary to turn the accident investigation over to another person or even an outside party. The ideal would be to have an investigation conducted by someone who is an expert in accident causation; experienced in investigative techniques; and fully knowledgeable of the work processes, procedures, persons, and industrial relations environment of the situation. Unfortunately, most of us live in the real world. Do not be afraid to call on any resources in your own firm.

9.13.2 Who and How Many People Should Investigate an Accident?

Depending on your work setting, an incident may be investigated by only the LSO, while in some settings both management and labor may have representation on the team.

9.13.3 Should the Immediate Supervisor be on the Team?

The advantage is that this person is likely to know the most about the work and persons involved and the current conditions. Furthermore, the supervisor can usually take immediate remedial action. The counter-argument is that there may be an attempt to gloss over the supervisor's shortcomings in the accident. This situation should not arise if the worker representative and the management members review all accident investigation reports critically.

9.13.4 How Do You Make Sure that Investigators Are Impartial?

An investigator who believes that accidents are caused by unsafe conditions will likely try to uncover conditions as causes. One who believes accidents are caused by unsafe acts will attempt to find the human errors that are causes. Therefore, it is necessary to examine briefly some underlying factors in the chain of events that ended in the accident.

9.13.5 Reason for the Investigation: Why Are We Doing This?

When accidents are investigated, the emphasis should be concentrated on finding the root cause of the accident rather than the investigation procedure itself. Once a laser accident or suspected incident has been reported to the facility LSO, swift action needs to be taken. First, work at the scene needs to stop. This is to determine what happened so corrective action steps can be taken to prevent it from occurring again. You are looking for a positive outcome; this is not a television crime scene where all parties are guilty. It is not uncommon to find that several people knew about the problem or the actions that caused the injury but took no action to see that it was corrected.

Written statements from the injured party and those in the area are important. Taking digital photos to help explain the scene and provide greater understanding is a great aid and time saver. Review written procedures if they exist as well as the training procedures of those involved. Did they understand the risk? Had they performed this activity before?

Gather evidence from many sources during an investigation. Get information from witnesses and reports as well as making direct observations. Interview witnesses as soon as possible after an accident. Inspect the accident site before any changes occur. Take photographs and make sketches of the accident scene. Record all pertinent data on maps. Get copies of all reports. Documents containing normal operating procedures, flow diagrams, maintenance charts, or reports of difficulties or abnormalities are particularly useful. Keep complete and accurate notes in a bound notebook. Record pre-accident conditions, the accident sequence, and postaccident conditions. In addition, document the locations of victims, witnesses, machinery, energy sources, and hazardous materials.

All parties should remember that most importantly, accident investigations are conducted to find out the causes of accidents and to prevent similar accidents in the future. They are not hunts for someone to blame the incident on (not finger-pointing exercises). Even laser incidents that involve no injury or property damage should still be investigated to determine the hazards that should be corrected. The same principles apply to a quick inquiry of a minor incident and to the more formal investigation of a serious event. As little time as possible should be lost between the moment of an accident and the beginning of the investigation. In this way, one is most likely to be able to observe the conditions as they were at the time, prevent disturbance of the setting, and identify witnesses.

The steps in accident investigation are simple: the accident investigators gather information, analyze it, draw conclusions, and make recommendations. Although the procedures are straightforward, each step can have its pitfalls. As mentioned above, an open mind is necessary in accident investigation. Preconceived notions may result in wrong paths being followed while leaving significant facts uncovered. All possible causes should be considered. Making notes of ideas as they occur is a good practice, but conclusions should not be drawn until all the information is gathered.

9.13.6 Summary of Investigation Steps

1. Stop work.
2. Get an understanding of the activity.
3. Take digital images.
4. Get statements from all involved.
5. Review existing procedures.
6. Review training records.
7. Follow-up:
   a. Develop corrective action plan, with user buy-in.
   b. Implement the plan.
   c. Prepare lessons learned and distribute them.
   d. Evaluate the effectiveness of the corrective action.

9.13.7 Physical Evidence

Before attempting to gather information, examine the site for a quick overview, take steps to preserve evidence, and identify all witnesses. Based on your knowledge of the work process, you may want to check items such as:

1. Positions of injured workers
2. Equipment being used
3. Materials being used
4. Safety devices in use
5. Position of appropriate guards
6. Position of controls of machinery
7. Damage to equipment

8. Housekeeping of area
9. Weather conditions
10. Lighting levels
11. Noise levels

9.13.8 Task

Here the actual work procedure being used at the time of the accident is explored. Members of the accident investigation team look for answers to questions such as:

1. Was a safe work procedure used? If not, why not?
2. Had conditions changed to make the normal procedure unsafe?
3. Were the appropriate tools and materials available?
4. Were they used?
5. Were safety devices working properly?
6. Was lockout used when necessary?
7. Was the worker distracted? If yes, why was the worker distracted?
8. Was the worker trained? If not, why not?

9.13.9 Equipment Issues

To seek out possible causes resulting from the equipment and materials used, investigators might ask:

1. Was there an equipment failure?
2. What caused it to fail?
3. Was the machinery poorly designed?
4. Were hazardous substances involved?
5. Were they clearly identified?
6. Was a less hazardous alternative substance possible and available?
7. Was the raw material substandard in some way?
8. Should personal protective equipment (PPE) have been used?
9. Was the PPE used?

9.13.10 Environmental Factors

1. What were the weather conditions?
2. Was poor housekeeping a problem?
3. Was it too hot or too cold?
4. Was noise a problem?
5. Was there adequate light?
6. Were toxic or hazardous gases, dusts, or fumes present?

9.13.11 Human Factors

1. Were workers experienced in the work being done?
2. Had they been adequately trained?

3. Can they physically do the work?
4. What was the status of their health?
5. Were they tired?
6. Were they under stress (work or personal)?

9.13.12 Management

The investigation team or person has to feel confident in their role to look at management's role in the accident. Management holds the legal responsibility for the safety of the workplace, and therefore the role of supervisors and higher management must always be considered in an accident investigation. Answers to any of the preceding types of questions logically lead to further questions such as:

1. Were safety rules communicated to and understood by all employees?
2. Were written procedures available?
3. Were they enforced?
4. Was there adequate supervision?
5. Were workers trained to do the work?
6. Had hazards been previously identified?
7. Had procedures been developed to overcome them?
8. Was regular maintenance of equipment carried out?
9. Were regular safety inspections carried out?

9.13.13 Gather Background Information

Often an overlooked source of information can be found in documents such as technical data sheets, maintenance reports, past accident reports, formalized safe-work procedures, and training reports. Any pertinent information should be studied to see what might have happened and what changes might be recommended to prevent recurrence of similar accidents.

9.13.14 Eyewitness Accounts

Witnesses should be interviewed as soon as it is practical after the accident. If witnesses have an opportunity to discuss the event among themselves, individual perceptions may be lost in the normal process of accepting a consensus view where doubt exists about the facts.

Although there may be occasions when you are unable to do so, every effort should be made to interview witnesses. In some situations witnesses may be your primary source of information because you may be called upon to investigate an accident without being able to examine the scene immediately after the event. Because witnesses may be under severe emotional stress or afraid to completely open up for fear of recrimination, interviewing witnesses is probably the hardest task facing an investigator.

Witnesses should be interviewed alone, rather than in a group. You may decide to interview a witness at the scene of the accident where it is easier to establish the positions of each person involved and to obtain a description of the events. However, it may be preferable to carry out interviews in the quiet of an office where there will be fewer distractions. The decision may depend in part on the nature of the accident and the mental state of the witnesses.

9.13.14.1 Interviewing: Getting a Story or Facts

Interviewing is an art that cannot be given justice in a brief document such as this, but a few dos and don'ts can be mentioned. The purpose of the interview is to establish an understanding with the witness and to obtain his or her own words describing the event.

9.13.14.2 Interview Details

1. Get preliminary statements as soon as possible from all witnesses.
2. Arrange for a convenient time and place to talk to each witness.
3. Explain the purpose of the investigation (accident prevention) and put each witness at ease.
4. Listen, let each witness speak freely, and be courteous and considerate.
5. Take notes without distracting the witness. Use a tape recorder only with the consent of the witness.
6. Use sketches and diagrams to help the witness.
7. Emphasize areas of direct observation. Label hearsay accordingly.
8. Be sincere and do not argue with the witness.
9. Record the exact words used by the witness to describe each observation. Do not put words into a witness' mouth.
10. Word each question carefully and be sure the witness understands.
11. Identify the qualifications of each witness (name, address, occupation, years of experience, etc.).
12. Supply each witness with a copy of his or her statements. Signed statements are desirable.

Do	Do Not
Put witness at ease	Intimidate the witness
Explain you want to determine what happened and why	
Let them talk and listen	Interrupt or prompt
Try to sense underlying feelings	Show your own feelings or emotions
Take short notes	Take lengthy notes while the witness is talking
Confirm that you have all statements correct	

9.13.14.3 What Type of Questions to Ask?

Ask open-ended questions that cannot be answered by simply "yes" or "no." The questions you ask the witness will naturally vary with each accident, but some general questions should be asked each time:

1. Where were you at the time of the accident?
2. What were you doing at the time?
3. What did you see and hear?
4. What were the environmental conditions (weather, light, noise, etc.) at the time?
5. What was (were) the injured worker(s) doing at the time?
6. In your opinion, what caused the accident?
7. How might similar accidents be prevented in the future?

9.14 WRITTEN REPORT: NOT FINISHED UNTIL THE PAPERWORK IS DONE

The following outline has been found especially useful in developing the information to be included in the formal report:

1. Background Information
   a. Where and when the accident occurred
   b. Who and what were involved
   c. Operating personnel and other witnesses
2. Account of the accident (what happened?)
   a. Sequence of events
   b. Extent of damage
   c. Accident type
   d. Agency or source of energy or hazardous material
3. Discussion (analysis of the accident: how and why?)
   a. Direct causes (energy sources, hazardous materials)
   b. Indirect causes (unsafe acts and conditions)
   c. Basic causes (management policies, personal or environmental factors)
4. Recommendations (to prevent a recurrence) for immediate and long-range action:
   a. Basic causes
   b. Indirect causes
   c. Direct causes (such as reduced quantities or protective equipment or structures)

An inquiry that answers these and related questions will probably reveal conditions that are more open to correction than attempts to prevent carelessness.

9.15 WHY SHOULD RECOMMENDATIONS BE MADE?

The most important final step is to come up with a set of well-considered recommendations designed to prevent recurrences of similar accidents. Once you are knowledgeable about the work processes involved and the overall situation in your organization, it should not be too difficult to come up with realistic recommendations. Resist the temptation to make only general recommendations to save time and effort.

For example, you have determined that a beam block that was knocked over contributed to an accident. Rather than just recommending "show more care around beam blocks," it would be better to suggest:

1. Anchor all beam blocks to the optical table.
2. Color beam blocks for easier recognition if out of place or tipped over.

Never make recommendations about disciplining a person or persons who may have been at fault. This would not only be counter to the real purpose of the investigation, but would jeopardize the chances for a free flow of information in future accident investigations.

In the unlikely event that you have not been able to determine the causes of an accident with any certainty, you probably still have uncovered safety weaknesses in the operation. It is appropriate that recommendations be made to correct these deficiencies.

9.16 WHAT SHOULD BE DONE IF THE INVESTIGATION REVEALS HUMAN ERROR?

A difficulty that has bothered many investigators is not wanting to lay blame. However, when a thorough worksite accident investigation reveals that some person or persons among management, supervisors, or workers were apparently at fault, then this should be pointed out. The intention is to remedy the situation, not to discipline an individual. Failing to point out human failings that contributed to an accident will not only downgrade the quality of the investigation; it will also allow future accidents to result from similar causes because they have not been addressed.

9.17 NOT EVERY LASER ACCIDENT INVOLVES AN EYE OR SKIN INJURY

Having a laser beam or reflection show up where it is not expected can also be the cause of an accident. The worker who is flashed with a construction laser and is startled may fall off the ladder he or she is standing on. In a laboratory somebody may drop a piece of expensive equipment. In either case the person operating the laser may now be in danger from the individual struck by the laser. The overwhelming number of accident reports from exposure to laser pointers fall into this group. Rarely has a laser pointer been the cause of a retinal eye injury, but the startle reaction has caused car accidents.

The most common effects from direct exposure to viewing the beam from a laser pointer are afterimage, flash blindness, and glare.

Afterimage is the perception of spots in the field of vision. This can be distracting and annoying, and may last several minutes, although there have been reports of afterimages lasting several days. Flashblindness is temporary vision impairment after viewing a bright light. This is similar to looking directly at a flashbulb when having a picture taken. The impairment may last several minutes. However, a longer look can cause serious damage to your eyes. It is worse if the laser beam is being projected through a piece of optical equipment, such as a telescope or a pair of binoculars. In these situations, the laser beam could actually burn a tiny spot, or cut open a blood vessel on the retina at the back of your eye. In a worst-case scenario, you could go blind.

Glare is a reduction or complete loss of visibility in the central field of vision during exposure to the direct or scattered beam. This is similar to viewing oncoming headlights on a dark night. Once the beam is out of the field of vision, the glare ceases. While this does not pose a hazard to the eye, it be a serious distraction. Glare can be exacerbated when the beam is reflected from a mirror-like surface.

9.18 TYPICAL LASER POINTER INCIDENTS

1. Law enforcement officers have reportedly drawn their weapons when the light from laser pointers is mistaken for a gun sight.
2. Laser beams projected into airspace to intercept aircraft have caused distractions and temporary vision impairment to pilots.
3. The operator of a roller coaster ride in a Hershey, Pennsylvania, amusement park claimed a laser flash temporarily blinded him.
4. Some sport venues and theaters have banned laser pointers after incidents with laser pointer beams.
5. A high school cheerleader reported being exposed at least three times. After the last episode, she reported first seeing green, then experiencing partial vision loss, which lasted for several months. An ophthalmic exam found no retinal damage. The cheerleader's vision returned to normal after several months.

9.19 GUIDANCE FOR LASER POINTER USE

Laser pointers are not toys. Use them with caution, and only for their intended purpose:

1. Never point a laser beam at anyone and never look directly into the beam yourself.
2. Never aim a laser pointer at surfaces that would reflect the light back, such as mirrors or mirrored surfaces, including fish tanks and windows.
3. Never leave a laser pointer where children might get their hands on it.
4. Choose a laser pointer that stays on only when you apply pressure with your fingers.
5. When you buy a laser pointer, choose one that has a clear warning on the label about the potential to cause eye damage. Read the instructions carefully and follow them closely.

9.20 EXAMPLES OF LASER ACCIDENTS

This section is a collection of documented laser incidents. Some took place in a medical setting and others in research and academic settings. Each, regardless of the nature of the injury, demonstrates that laser accidents happen and the user must be aware.

One of the oldest laser accident databases was originated and kept by Rockwell Laser Industries (RLI), a laser-safety consulting firm that collects documented accident accounts. Approximately every 10 years they publish the updated totals of their database. The firm's Web page contains a form where accidents and incidents can be reported, which can be found at http://www.rli.com/.

The Department of Energy (DOE) has an extensive accident-reporting database, called the Occurrence Reporting and Processing System (ORPS). Many kinds of accidents and incidents are tracked through this system. Over the past 5 years, a number of the national laboratories have reported a laser injury or a break in laser safety procedures. A national laboratory lessons learned Web page can be found at http://tis.eh.doe.gov/paa/oesummary.

The U.S. military keeps databases of incidents and injuries. One is the Laser Accident and Incident Registry, U.S. Army Medical Research Detachment, Walter Reed Army Institute of Research.

9.20.1 Medical Device Report (MDR) Report Key 397304, Report Number 1218402-2002-00019

Event description: The representative of a medical laser manufacturer sent a user an extra pair of laser protective eyewear. The problem was that the extra pair was not for the wavelengths produced by the laser. The eyewear was correctly marked with wavelength OD, but the doctor who received it did not check. He received a reflection that passed through the eyewear and caused an eye injury.

9.20.2 Access Number M320116, MDR Database

Event description: After the kit and laser were connected, it was observed that the red HeNe aiming beam was not visible at the laser kit tip. The physician elected to proceed with the procedure. When the laser control foot switch was depressed, no laser energy was emitted from the laser tip. Flames appeared on the sterile drape where the laser fiber had been resting. No one was injured. Analysis of the returned product pointed to a broken laser fiber, probably caused by mishandling.

9.20.3 Access Number M751889, MDR Database

Event description: A male patient was undergoing microlaryngoscope laser surgery to remove a nodule from his vocal cord. The endotracheal tube was in use when an ignition occurred. The patient sustained severe burns to his vocal cords and surrounding tissues, causing obstructive swelling.

9.20.4 Access Number M817394, MDR Database

A nurse in the operating room leaned into the laser system at the point where the laser delivery fiber attached to the laser console and caused the fiber to break at the connector. The broken fiber caused a burn in the nurse's clothing, penetrated the clothing, and caused a small nonserious burn on the nurse's abdomen.

9.20.4.1 Post Doc Incident

A postdoctoral student was attempting to align an unfocused laser beam when a stray beam from an optic polarizer he was holding glanced into his face. No protective eyewear was worn, and eye pain developed in 24 hours.

9.20.5 ORPS Report CH-AA-ANLE-ANLEER-1999-0005

A researcher performing welding of 24-inch-long aluminum plates using 4 kWatt of power from a carbon dioxide laser caused a ceiling panel to start burning, which was detected when the room's smoke detector went off.

9.20.5.1 Experience is No Guarantee of Safety

The person injured had 15 years experience with lasers. The experiment was running 1 mJ, 500 HZ, and femtosecond pulse length, with a beam size of several centimeters. The beam was aimed in an upward direction toward a periscope. The beam output was not lowered, because it burned through the neutral density filters. The two researchers decided that if they were careful, it would be all right to insert a mirror into the full-power beam path. This activity was a violation of written procedures for the experiment. An IR viewer was not used. One researcher was placing the mirror into the beam path and was struck by reflection from the corner of the mirror. The person heard a popping sound from his eye, followed by swelling of the eye. The result was a 100-micron spot injury. Vision went from 20/50 to near blindness; the researcher still cannot read large print.

The chemistry division, where this incident occurred, houses nearly 50% of all lasers at Argonne. All laser work stopped for a month while all groups were required to review laser use procedures or generate them if they had none. The chemistry division halted all laser work for more than 30 days. The researcher involved was banned from laser work for 30 days.

9.20.6 From the Military Database

9.20.6.1 GLIN 4

The subject experienced a large flash of light and described seeing black circles and experiencing an almost complete loss of vision. The vision slowly cleared in the next 48 hours, but many dark brown spots continued to form a "curtain" in front of the eye. Assuming the pupil diameter was about 7 mm (dark environment), the retina was exposed to about 4.3 mJ of total beam energy. The resulting pressure wave damaged all the structures in the image area. The bruch's membrane was ruptured posteriorly, the sensory retina and nerve fiber layer were disrupted anteriorly, and at the retinal level, the photoreceptors sustained severe damage peripherally.

9.20.6.2 Q-Switched Nd:YAG Injury (Israel)

The injury involved a paramacular burn, later puckering, and permanent loss of visual acuity. The laser had a 1064-nm wavelength with a 20-nsec pulse duration, 20-mJ/cm squared beam energy, and a 4.3-mJ/pulse duration exposure (assuming a 7-mm pupil). The distance to the laser was 0.03 m.

9.20.6.3 Military Range Finder

A civilian in Germany purchased a British military range finder at a local flea market. He then accidentally exposed his right eye, causing a central cub and intraretinal hemorrhage. Vision was 20/100. At 4 weeks the eye was clear with a pigmented spot 200 mm from the fovea. Aggressive treatment consisted of intraocular pressure control and medications and staying face down for 5 days. The Nd:YAG laser was 1060 nm, with 10 mJ at 10-nsec pulses.

9.20.6.4 Airline Pilot

The pilot of a United Airlines 737 aircraft was tracked for 3 sec with a green light thought to be an argon laser. The aircraft was flying at 13,000 feet at 21:10 hours. The event occurred April 16, 1997. The pilot landed the aircraft safely. The beam seemed to come from a residential area. An afterimage persisted for some time.

9.20.6.5 Airline Pilot

In October 1996, a Southwest Airlines flight was taking off from Las Vegas, Nevada. At an altitude of 17,000 feet, the beam from a nearby hotel-based laser light show entered the cockpit and caused the pilot to be temporarily blinded in the right eye. The pilot turned control of the plane over to the copilot. An argon laser is assumed to be the laser radiation source.

9.20.6.6 Student Sues

The following example provides a major argument supporting "real" laser safety policies and practices and why these should be a significant part of all laser programs. A research assistant received exposure in both eyes while working unsupervised in the lab. She was in the lab reportedly having an argument with her boyfriend, who was not authorized to be there. The assistant was not wearing laser protective eyewear, although it was available. She claimed that the professor never wore the eyewear and that it was just a "paper policy."

Following the event, the student sued the university for $39 million and reportedly settled for $1 million out-of-court. She claimed the protective eyewear and laser safety policy was only on paper and was not practiced. She ignored a policy that *no* guests were allowed in the labs and that supervision was required to operate the laser. Nonetheless, she collected on the claim. Such a settlement clearly indicates that the courts tend to favor the injured party. Obviously laser safety programs that are viewed only as "paper policies" can ultimately be very expensive.

9.20.7 Data from the NASA Aviation Safety Reporting System Database

There are many examples of commercial flights in which the pilots suffered eye damage from lasers. These include aircraft landings at Honolulu, Las Vegas, Miami, New York, Los Angeles, and Phoenix. In Phoenix, one crew member was flash-blinded, with resulting afterimages and loss of night vision for about 1½ hours. Takeoffs have also been affected: in a 737 outbound from Los Angeles, two pilots were struck by a blinding flash that lasted 5 to 10 sec. The first officer had burns on the outer eye and broken blood vessels. In a flight from Cleveland, one crew member received a bright blue light in his right eye and experienced vision impairment for the next 1½ hours. Data from the National Air Intelligence Center indicate that, in the United States alone, commercial lasers have caused over 50 blinding incidents. Lasers have also injured a number of air force personnel. For example, the Palace Casino's laser show laser-illuminated a C-130 landing at Keesler Air Force Base. The flight engineer, who was looking straight ahead, was blinded for 3 to 5 sec and then experienced blurred vision. The next day, he experienced eye pain requiring eye drops. In April 2005 two Royal Canadian Air Force helicopter pilots were laser-illuminated from a Russian trawler during a routine mission.

9.20.7.1 Severe Skin Exposure from a Xenon Chloride Excimer Laser

A male research technician was working on a class I XeCl excimer laser operating at a wavelength of 308 nm when the protective enclosure was opened near the pre-ionization discharge. He was exposed with several reflected laser pulses while the laser was operating in the pulsed mode at approximately one pulse per second. The technician had removed a portion of the protective housing that covered a beam splitter and was watching for an abnormal electrical discharge inside the laser chamber. About 15 mJ per pulse reflected off the beam splitter and produced four distinct burns on his neck. As this wavelength, no heat was detected and the burns were not evident until several hours after the exposure. The beam reflections from the beam splitter that exposed the technician's neck area were unfocused. He felt nothing at the time, but several hours later, four burns (probably photochemical in origin) appeared on his neck. These areas required nearly three weeks to heal. The technician was using eye protection.

9.20.7.2 9-mm Murphy Style Cuffed Endotracheal Tube

The patient was in the intensive care unit with multiple medical problems including uncontrollable tracheal hemorrhage. On December 10, 1991, the patient was taken to the operating room for a rigid bronchoscopy with a laser. The endotracheal tube that was in place was removed and a number 9 PVC tube was placed. The procedure was then changed to a fiberoptic bronchoscopy with a YAG laser. The fiberoptic bronchoscope was positioned approximately 1 cm distal to the endotracheal tube, and the laser was situated about 1 cm distal to the tip of the fiberoptic bronchoscope. The power was set at 30 W with 1-sec pulses. After four or five doses of the laser coagulation therapy, the tip of the endotracheal tube ignited and a tracheal fire occurred. The patient suffered thermal burns of the trachea, left main stem bronchus, left upper lobe bronchus, and left lower lobe bronchus. This patient was gravely ill. It was necessary during the procedure to deliver 100% oxygen. When the normal procedure is performed, the amount of oxygen is approximately 30%. The patient died 9 days following the incident. No sample was returned to the company from the hospital. The hospital's director of risk management explained that the hospital does not blame the endotracheal tube for the trachea fire; it was a risk they had to take to attempt to prolong the life of a gravely ill patient. Labeling instructions warn against using a laser or electrosurgical device in conjunction with the endotracheal tube because contact with the tube could result in its rapid combustion.

9.20.7.3 Gastroenterology-Urology

As the device was being used with a CO_2-cooled laser fiber in the intra-abdominal area, the patient underwent cardiac arrest. It was alleged that an air embolism occurred. The patient was resuscitated and no permanent injury occurred. A laparoscopy was performed on a 32-year-old woman. The procedure went smoothly, was successful, and nothing unusual was noticed. At the completion of the procedure, the anesthesiologist noticed that the patient had an electrocardiogram pattern but no pulse. It was determined that the patient had a CO_2 gas embolism in the heart. The intra-abdominal pressure was 30 mm Hg. The surgeon immediately released CO_2 gas from the abdomen.

The hospital has not conclusively determined what caused the embolism. The laser fiber continuously emits CO_2 gas to cool the tip. There was some bleeding at the treatment site, but it was not excessive. The laser was used for ablation and coagulation. The hospital is not sure if the laser fiber may have directly injected CO_2 into an open blood vessel. The high intra-abdominal pressure may have forced gas into the open circulatory system. The hospital reports that the laser has no gas shut-off or pressure-sensing features. The insulator has a red light, which comes on when the abdominal pressure exceeds the preset pressure by 4 mm Hg. The insulator interrupts CO_2 gas flow at this pressure. In addition, an alarm sounds at sustained pressures over 20 mm Hg. No one heard an alarm, but it was reported that the laser was "very noisy." The insulator was returned to the manufacturer for inspection. The manufacturer concluded after testing the instrument that the device did not malfunction and that the incident is the result of an oversight on the part of the surgical team and the fact that the CO_2 gas feeding the laser fiber had no feedback system to control the CO_2 flow.

In another incident the laser appears to have continued to work after the footswitch was released. The affected laser system was returned to the manufacturer for investigation, but they could not duplicate the claimed malfunction. The manufacturer found that the laser system performed to specifications. There is triple redundancy in the design. Three failures would need to occur simultaneously: (a) the laser would fail to turn off, (b) the shutter would fail to close, and (c) the mechanism and software that detect shutter failures would fail. The manufacturer believed that the reported malfunction did not relate in any way to the malfunction that led to the recall of the device (recall number z-164-3). This recall was based on the possibility of a shutter coming loose and causing an unintentional radiation exposure during power-up or power seeks. The reported malfunction here is that the laser continued lasing after the release of the foot pedal. The shutter becoming loose would lead to the laser malfunctioning. In addition, release of the foot pedal would signal the laser to stop lasing. The manufacturer made a follow-up phone call to the doctor's office. The doctor reported continuing to hear a noise from the laser system after releasing the foot pedal. The company representative explained that the air pump in the laser system operates for approximately 2 sec after the foot pedal is released. This is normal operation. The doctor now believes that he confused the air pump with the continued lasing. The doctor also used thermal paper to verify that the laser was not continuing to emit energy. He said there were no burn spots on the tissue.

Another complaint was that a laser system continued to lase for approximately 1 sec after the foot switch was released. The affected laser system was returned to the manufacturer for investigation. The claimed malfunction could not be duplicated. The laser system performed to specifications (radiation ceased within 100 msec). There is triple redundancy in the design. Three failures would need to occur simultaneously as in the previous example. The manufacturer believes that the malfunction does not relate in any way to the malfunction which led to the recall of the device (recall number z-164-3). The recall was based on the possibility of a shutter coming loose and causing an unintentional radiation exposure during power-up or power seek. The reported malfunction here is that the laser continues lasing after release of the foot pedal. The shutter becoming loose would lead to the laser faulting. In addition, release of the foot pedal would signal the laser to stop lasing. The manufacturer made a follow-up phone call to ask the doctor what the indications of continued lasing were. The doctor responded that there was charring, not just bubbles or the sound of the air pump. Lasing did not occur for more than half a second. The laser was being operated in mode 7, 2 to 3 W (pulse mode).

9.20.7.4 Iris Medical Slit Lamp, Product Safety Failure

A sales representative demonstrated an iris medical occu-light glass laser system with a slit lamp adapter delivery device to three doctors at a hospital. While making practice burns on a business card, doctor 2 commented on the brightness through the oculars after test firing the laser. Doctors 2 and 3 said they saw spots after test firing; doctor 1 and the sales representative noticed no unusual brightness. Upon further inspection of the slit lamp adapter delivery device, it was noticed that the safety filter frame was labeled 810 nm. The wavelength of the laser being demonstrated was 532 nm. On May 6, 1999, it was verbally reported by the hospital risk management department to the sales representative that doctor 3 was found to have 20/50 vision and three suspected laser burns on his retina. The ophthalmologist who examined doctor 3 indicated that it could take several months for the vision to recover and stabilize. Doctors 1 and 2, as well as the sales representative, reported no effects. The exact laser powers and exposure durations used during these exposures have not been determined, but it appears that the laser power did not exceed 300 mW and the duration was set to 50 msec.

9.20.7.5 Ophthalmic Excimer Laser

An ophthalmologist arranged for his patients to travel for the purpose of eye surgery by excimer laser. With two doctors operating the laser, the patient's right eye was contacted by the laser beam 1 cm off center. The procedure was terminated at that point, but the patient's vision was changed from –5.50 to –8.50. The doctor stated that this had happened previously with another patient. The doctor added that it had something to do with the gears in the control mechanism, but that no one knew why it happened. The manufacturer's representative asked the doctor why he would use such an unreliable machine for this work, and the doctor said that it was a rare occurrence that he likened to driving his car. The doctor added that it would not be high on his list of things to worry about. Today, over a year later, the patient's vision is –7.50 in the right eye.

9.20.7.6 Holmium:Yttrium Aluminum Garnet (YAG) Laser System

When the doctor had finished lasing, the laser continued to lase, even though the foot switch was deactivated (released). Lasing was terminated via the emergency off button. No deaths or serious injuries occurred. The laser was thoroughly checked out a short while later. The problem could not be reproduced. At the time of the problem, a large amount of liquid was noted around the foot switch. Preliminary investigation into the event and simulated lab testing tentatively identified the event as being attributable to shorted connections in the foot switch caused by excessive liquid in the foot switch assembly.

9.20.8 Fire Case #1

After the kit and laser were connected, it was observed that the red HeNe aiming beam was not visible at the laser kit tip. The physician elected to proceed with the procedure. When the laser control foot switch was depressed, no laser energy was emitted from the laser tip. Flames appeared on the sterile drape where the laser fiber had been resting, but they were quickly extinguished. No one was injured. Analysis of the returned product pointed to a broken laser fiber, probably caused by mishandling.

9.20.9 Fire Case #2

A nurse was readjusting the laser system for the next patient. During the verification of the energy, she heard an unusual noise. The nurse put the system in "standby" mode and sought assistance from her supervisor. When the supervisor took the system from "standby" mode to "ready" mode, she heard a loud bang and subsequently smelled smoke. Hospital personnel extinguished a small fire, which was contained within the laser system. An internal high-voltage arc ignited the component's supporting plastic material, resulting in the internal fire. In order to prevent this from recurring, a design change that increased the high-voltage clearances and changes in associated materials was made.

9.20.10 Fire Case #3

An incident occurred as a result of a laser-assisted turp procedure. Fiber from one manufacturer was utilized with another manufacturer's laser system. The physician had initiated the procedure and had lased three times for approximately 60 sec each time when the laser nurse smelled smoke and reported that her clothing was in flames. A hole in the clothing approximately $3 \times 2"$ was created by the burn. No injuries occurred to the patient, surgeon, or nurse. The manufacturer's engineers visually examined the fiber. The failure occurred in the center of the small connector, and the proximal aperture was not damaged. Since all fibers are thoroughly tested prior to shipping and the hospital reported that this fiber worked initially, the manufacturer's conclusion was that the beam output from the laser system was not compatible with the other manufacturer's fiber. This resulted in a hot spot inside the fiber that caused the fiber to fracture and fail. The fiber was not intended for use with lasers producing a beam divergence from the laser system's aperture greater than $17°$ full angle.

9.20.11 Lubricant Fire

USP recently received a report regarding the use of a petroleum-based eye lubricant, Lacri-Lube® S.O.P.®, manufactured by Allergan. In the reported incident, the lubricant was applied to the eyelid and periorbital area of a 4-year-old patient undergoing laser surgery to remove warts around the eye. When a laser was directed to the area, the eyelashes and eyelid ignited, injuring the patient's left eye.

Lacri-Lube S.O.P. consists primarily of white petrolatum and mineral oil, with small amounts of chlorobutanol and lanolin alcohols. According to the reporter, the hospital where this incident took place is replacing oil-based ocular lubricants with water-based products for procedures in which a laser or other heat source is used in close proximity to the lubricant.

9.20.12 Range Finder

A 24-year-old male patient visited an emergency department with a history of decreased vision in his right eye after looking into the laser beam of a range finder one hour previously. The findings were as follows: visual acuity OD sc=0,2 on the anterior segment without damage. Biomicroscopy revealed a retinal lesion located 200 µm parafoveal, with subsequent submacular bleeding. This hemorrhage had spread below the fovea and was subretinal as well as intraretinal. Infrared laser scanning ophthalmoscopy showed a stretched outer vitreous membrane. A technical examination of the range finder revealed a pulsed Nd:YAG laser of 1064-nm wavelength, 10-nsec pulse duration, 10-mJ pulse energy, and a beam diameter of 2.5 cm. Treatment consisted of 100 mg prednisolone for 9 days systemically, and 50 µg rTPA was injected into the vitreous under local anaesthesia. Furthermore, 0.5 ml pure C2F6 was injected via pars plana 24 hours later. Postoperative follow-up will be demonstrated.

Macular injuries caused by range finders cause substantial impairment. There is a gray market for military laser equipment that is not only illegal but prevents any safety instruction on apparent dangers.

9.20.13 Bystander Observes Diffuse Beam During Laser Operation

An employee returning from a break heard and investigated a hissing sound coming from an equipment area outside a research facility that contained a liquid nitrogen tank. Upon further investigation of this sound, the employee observed a green flashing light through a 12" exterior wall opening. The employee reported the incident and sought medical evaluation 2 hours later.

The incident was caused by a research scientist who had used two class IV Nd:YAG lasers to visualize airflow in a test chamber and removed an interior metal cover from this opening for compressed air relief from the test chamber. A second metal cover, attached to the building outside this opening, was observed lying on the ground. Based on the laser operating parameters, the diffuse reflection did not exceed the MPE at the employee viewing location, and a retinal examination of the employee confirmed no injury. However, the NHZ from this diffuse reflection did extend several feet beyond the exterior wall and could have resulted in a retinal injury to the employee or a bystander if the reflection was observed at a smaller viewing distance.

9.20.14 Other Laser Accidents

1. An Nd:YAG laser beam of 1060 nm and 1.2 kW peak power, at 1 psec and 100 MZ, was directed across a room to a table where a sample bottle was located. The beam was larger than the bottle, and about 10% of the beam went into a nearby area where a graduate student was working. The student heard a popping sound when the beam struck his retina. A white macular damage zone was confirmed by examination.
2. A technician inserted a polished test slide into a Ti:Sapphire laser beam (806 nm, 10 to 50 uJ/pulse at 10 Hz) to determine whether the beam would harm a detector surface. The beam reflected form the test slide into his eye. A flash was observed and was followed by an ocular hemorrhage, and blood filled his eye. He experienced no pain or shock. At 3 weeks postexposure, there was no vision restoration in the eye.
3. Viewing a laser at a distance of 3 km with a 7× telescope, a soldier experienced a bright orange-red flash, which caused a hemorrhage in the left eye. The source was a ruby laser at 694 nm. At 10 days postexposure, the retinal burn measured 150 μm size with a 1- to 2-mm hemorrhage zone. Acuity was measured at 10/400. After 6 months the retina showed a 7° degree central scotoma with a 2° surrounding scotoma.
4. An engineer received third-degree burns on his back and upper right arm when a carbon dioxide beam from a 1-kW carbon dioxide laser was accidentally turned on. The improper restoration of circuit wires open to the laser control system was deemed to be the cause of the accidental discharge.

9.21 LASER ACCIDENT ACTION PLAN

While our efforts are directed toward preventing laser accidents, one needs to a have a plan to follow in case an accident happens. More importantly, laser users need to know what that plan is. One solution is to place in every laser work area a poster that provides the names and current phone numbers of emergency personnel and steps to be followed until help arrives. A posted page or section in a laser procedure manual is recommended. The sample poster shown in Figure 9.2 could also be prepared as a Web page. The Web page approach has the advantage of quick updates, such as phone numbers and personnel changes. The disadvantage is that it may not be thought of during the excitement of the moment.

Laser Eye Accident Emergency Procedure

If you suspect a laser eye exposure, complete the following:

1. Immediately call _____________
2. Notify your supervisor.
3. Notify others in the work area.
4. Do not drive yourself to the Health Services Department or a doctor.
5. Remain in a sitting position during transport and examination to prevent further damage to the retina.
6. Ensure that you receive an examination by an ophthalmologist to determine whether an injury exists.
7. Safe the laser and leave the scene unchanged.
8. LSO is to be notified.

FIGURE 9.2 Action poster for suspected laser eye injury.

9.21.1 General Guidance for Suspected Eye Injury

1. Keep the individual calm, preferably seated or lying down; avoiding panic or shock is the main goal.
2. Call for assistance.
3. If your institution has a central help number for emergencies, call that number. Otherwise have the number of the local trauma center available. Do not count on using or finding the phone book.
4. If you have a medical clinic on site, they can be called or notified by your response office. If medical or security is called first, they should have standing instructions to contact the LSO.
5. It is important that the medical facility have some understanding of laser eye injuries as well as the laser mechanism.
6. Transport the person to medical.
7. Many large facilities have a fire department or security force that can transport the individual; they should be instructed on how to handle a person with an eye injury.
8. Notify the individual's or the area supervisor, along with the LSO and others working in the same area or on the same equipment.

This information that can be included on a poster or Web page. Additional examples from three universities can be seen at the following Web pages:

1. http://blink.ucsd.edu/Blink/External/Topics/How_To/0,1260,12286,00.html?coming_from=Content
2. http://radsafe.berkeley.edu/lsr-appg.html
3. http://ehs.unc.edu/radiation/manual/laser/I0-1.html
4. http://www.safety.duke.edu/Radsafety/laser%5Fpolicy/#8_1

In case of laser accident with suspected eye injury:

1. Determine if any local assistance is available, such as a coworker.
   a. If yes, this person follows steps 2 and 3.
   b. If no, call for assistance. Do not go by yourself.
2. Keep the person as calm as possible.
3. Call the medical or fire department.
   a. Regular hours: you can transport the person to medical.
   b. Off hours: fire department to provide transportation.
4. Notify the individual's or the area supervisor.
5. Notify the LSO.
6. Work needs to stop until an evaluation is conducted to see if a systematic error exists.

In case of a suspected skin injury:

1. Keep the person calm.
2. Call the medical or fire department.
3. Follow the same steps as for an eye injury.
4. If the injury is a hand burn with no active bleeding, you can transport yourself to medical.
5. At the scene, reassurance is the most important thing to provide.
6. Not all laser injuries have an immediate affect on vision, so initial and follow-up eye examinations are critical.
7. Let others know what happened.
8. Work needs to stop until an evaluation is conducted to see if a systematic error exists.

9.22 MEDICAL FACILITY

Depending on your community, you may have limited choices on where to send an individual with a suspected laser eye injury. It is important that the facility have some understanding of laser eye injuries as well as the laser mechanism. Usually, once the injured individual informs the medical staff that he or she works with or around lasers, any injury, particularly retinal, will be assumed to be laser related. Medical personnel may overlook the fact that many other optical causes or diseases could be the reason for visual problems or defects. Provide the name of a retinal specialist to the injured individual for further evaluation or follow-up in cases involving visible or near-infrared laser radiation. Not all laser injuries have an immediate effect on vision; consequently, initial and follow-up eye examinations are critical (Figure 9.2).

9.23 CAN LASER ACCIDENTS BE PREVENTED?

Laser accidents and near misses happen all the time. Some are caused by user error and some by equipment failure, and most can be prevented. It really comes down to the desire of the laser users to prevent accidents, their technique and basic laser safety training, and an appreciation of how quickly an accident scenario can develop. The laser use community needs a cultural change. It is no longer acceptable for people to think that if they are injured it only affects them. We are all affected. We must be responsible for each other as well as ourselves.

BIBLIOGRAPHY

Rathkey, A.S., Accidental laser burn of the MACULA, Arch. Ophthalmol. 74, 346–348, 1965.

Blancard, P., et al., A propos d'une photocoagulation maculaire par laser, accidentelle, *Ann. Oculist*, 198, 263–264, 1965.

Jacobson, J.H. and McClean, J.M., Accidental laser retinal burns, *Arch. Ophthalmol.,* 74, 882, 1965.

Zweng, H.C., Accidental Q-switched laser lesion of human macula, *Arch. Ophthalmol.*, 78, 596–599, 1967.

Curtin, T.L. and Boyden, D.G., Reflected laser beam causing accidental burn of retina, *Am. J. Ophthalmol.*, 65, 188–189, 1968.

Litvin, M.S., et al., Burn injury after carbon dioxide laser radiation, *Arch. Surg.*, 98, 219–222, 1969.

Armstrong, C.E., Eye injuries in some modern radiation environments, *J. Am. Opt. Assn.*, 41, 55–62, 1970.

Henkes, H.E. and Zuidema, H., Accidental laser coagulation of the central fovea, *Ophthalmologica*, 171, 15–25, 1975.

Osipov, G.L. and Pyatin, M.M., Eye injury by laser beam, *Vestnil Oftal'mologii*, 1, 50–51, 1978.

Burgess, G.F. and LeJeune, F.E., Endotracheal tube ignition during laser surgery of the larynx, *Arch. Otolaryngol.*, 105, 561–562, 1979.

Sliney, D.H. and Wolbarsht, M., *Safety with Lasers and Other Optical Sources*, Plenum, New York, 1980, p. 103.

Prammer, G., Wiesflecker, J. and Grabner, G., Makulaver- letzung durch Festoorperlaser, *Klin. Mbl. Augenheilk*, 176, 326–327, 1980.

Boldrey, E.E., et al., Retinal injury due to industrial laser burns, *Ophthalmology*, 88, 101–107, 1981.

Balashevich, L.T., et al., Some incidents of eye damage by laser radiation, *Vestnil oftal'mologii*, 1, 60–61, 1981.

Volkov, V.V. and Balashevich, L.T., et al., Course and outcome of accidental laser injuries to fundus oculi, *Oftal'mologicheskiy Zhurnal*, 36, 3, 1981.

Bleckmann, H. and Zorn, R., Akzidentelle Laserkoagulation der Makula, *Klin. Mbl. Augenheilk*, 179, 38–40, 1981.

Gayday, V.M. and Filippenko, V.I., Accidental laser injury to fundus of both eyes, *Voyenno-Meditsinskiy Zhurnal*, 5, 56–58, 1982.

Glovinsky, Y., Regenbogen, L., Bartov, E., et al., Macular pucker following accidental laser burn, *Metab. Pediatr. Syst. Ophthalmol.*, 6, 355–359, 1982.

Goldberg, M.F., Young, R.S., Read J., et al., Macular hole caused by a 580 nm dye laser operating for 10 nsec., *Retina*, 3, 42–44, 1983.

Fowler, B.J., Accidental industrial laser burn of the macula, *Ann. Ophthalmol.*, 15, 481–483, 1983.

Asano, T., Accidental YAG laser burn, *Am. J. Ophthalmol.*, 98(1), 116–117, 1984.

Wolfe, J.A., Laser retinal injury, Mil. *Med.*, 150, 177–185, 1985.

Lang, G.K., Lang, G. and Naumann, G.O.H., Akzidentielle bilaterale asymmertrishe Rubin-Laser-Makulopathie, *Klin. Mbl. Augenheilk,* 186, 366–370, 1985.

Rockwell, R.J., Learning from case studies: how to avoid laser accidents, in *Clinical Lasers: Expert Strategies For Practical and Profitable Management*, Breedlove, B. and Schwartz, D., Eds., American Health Consultants, Atlanta, GA, 1985, chap. 4.

Improper safety glasses key factor in serious eye injury, Issue No. 11, (DOE/EH-0007), *Serious Accidents,* Assistant Secretary for Environmental Safety & Health, U.S. Dept. Energy, Washington, DC, January, 1986.

Rockwell, R.J., Alignment accidents, *Radiant Resources Newsletter*, Vol. VII, August, 1987.

Rockwell, R.J., Summary of medical laser incidents reported to FDA From 1983–1986, *Radiant Resources Newsletter*, Vol. VII, March/April, 1987.

Rockwell, R.J., *Laser Safety Training Manual,* 3rd ed., Rockwell Associates, Cincinnati, OH, 1988.

Bauman, N., Laser accidents: why only 10% get reported, *Laser Med. Surg. News Adv.*, August, 1–7. 1988.

Bandle, A.M. and Holyoak, B., Laser incidents, in *Medical Laser Safety: Report No. 48*, United Kingdom Institute of Physical Sciences, London, 1988, chap. 6.

Haifeng, L., et al. Ocular injuries from accidental laser exposure, *Health Phys.*, 56(5), 711–716, 1989.

Gabel, V.P., et al. Clinical observations of six cases of laser injury to the eye, *Health Phys.*, 56(5), 705–710, 1989.

Rockwell, R.J., Jr., Laser accidents: are they all reported and what can be learned from them? *J. Laser Appl.*, October, 53–57, 1989.

Tomita Y., et al., Mutagenicity of smoke condensates induced by co_2 laser irradiation, *Mutat. Res.*, 89, 145–149, 1981.

American National Standards Institute, *American National Standard for the Safe Use of Lasers: ANSI Z-136.1 (1993)*, Laser Institute of America, Orlando, FL, 1993.

Rockwell, R.J., Jr., Laser accidents: reviewing thirty years of incidents: what are the concerns — old and new. *J. Laser Appl.*, 6, 203–211, 1994.

10 An Explanation of Control Measures

And the standards say thou shall & should ...

10.1 ADMINISTRATIVE CONTROLS

While it would be ideal to always have laser products that can be used, maintained, and serviced without us having to think about the safety issues, this can rarely be achieved in practice. Even a laser printer, that most ubiquitous of office tools, can present a risk to engineers during servicing — and indeed during routine maintenance operations. The risks are unlikely to be due to the laser beam for this application, but certainly some components can be very hot. Our protection is provided through what are termed administrative control measures. Essentially, these are control measures that are not engineered into the product and are not personal protective equipment; it is a catch all of the "other" control measures.

Administrative control measures include training, instructions, and warning signs. The important aspect of administrative control measures is that they are only effective if people act on them. Therefore, they tend not to be fail-safe. For this reason, we should always consider engineering control measures first.

10.2 FOR THE INSTITUTION

For the institution, the most important administrative control measure is the laser-safety management program. This is the process through which you decide where you are on your road to successful laser safety management and how you will get where you want to (or should) be. The institution should have a clear laser safety policy and a laser safety officer (LSO).

10.2.1 Policy

The laser safety policy is a major component of the finish line on your road to successful laser safety management. It is a statement of where you want to (or should) be. It may also outline a methodology for getting there. At its simplest, the laser safety policy could state the following:

> It is our policy that all laser safety products used within the organization be class 1 products during normal use, maintenance, and service operations. As such, no employees or others should be at risk of injury from the use of the laser product. All nonbeam hazards should be controlled by engineering means so that the risk is negligible.

This may be workable for a few laser products in some applications. However, the policy becomes meaningless if it is just a statement on paper that means little in practice. What if a class 2 laser pointer is regularly used for presentations to customers? You are not complying with your own policy. Some national regulators take the view that failure to comply with your own written systems is as bad as not complying with legislation — and they may prosecute you.

Thus, the message is to write a policy that reflects what you can reasonably achieve, stick to it, and prove that you are sticking to it. The organization may have an overall safety policy statement. It may be that this statement adequately covers laser safety. If so, it is worth stating so, probably in the laser-safety management program.

A more effective laser safety policy may be as follows:

> The organization aims to provide a safe environment for its employees and others. Where practicable, only class 1 laser products should be used on-site. The classification of the product should also be considered during user maintenance and servicing carried out by engineers. Where products of a class higher than class 1 are to be used, or where the assumptions for class 1 are compromised, these should be subject to a formal assessment in conjunction with the laser safety officer [or whoever is appropriate] before the product is purchased. The conclusions from the assessment should be recorded in the laser-safety management program for the laser application. All laser products should comply with the requirements of the laser safety standard current at the time of acquisition. This policy applies to laser products, whether they are purchased, hired, loaned, or demonstrated on-site. Nonbeam hazards must also be considered part of the assessment for any laser product coming on-site and the results of the assessment should be recorded in the laser-safety management program for the laser application.

After you have produced a laser safety policy, consideration then has to be given to how this is communicated to everyone who needs to know. There is no point in having a policy that is not known to likely laser users and, perhaps as important, the equipment acquisition section of the organization.

10.3 RESEARCH LAB ADMINISTRATIVE CONTROLS

The important aspect and shortcoming of administrative controls is that they are only effective if people act on them. Therefore, they tend not to be fail-safe. This is why engineering controls should always be given first consideration. How appropriate an administrative control is depends on how well people use it and the risk factor of the control to be mitigated. The laser classification scheme provides a good indication of laser beam hazard.

Laser safety in the research lab is achieved by a combination of engineering and administrative controls. The most important administrative control is the standard operating procedure (SOP), which was described in detail in Chapter 4. Following is a list of standard-based administrative controls, followed by an explanation of each:

1. Standard operating procedures
2. Education and training
3. Authorized personnel
4. Alignment procedures
5. Protective equipment
   - Eye protection
   - Protective windows
   - Protective barriers and curtains
6. Designation of areas
   - Signs
   - Labels
7. Spectators
8. Service personnel
9. Laser optical fiber transmission systems
10. Laser robotic installations
11. Skin protection

10.3.1 Standard Operating Procedures

The LSO needs to require and approve written standard operating, maintenance, and service procedures for class 3B and 4 lasers and laser systems.

10.3.2 Education and Training

Education and training must be provided for operators, maintenance, and service personnel for class 3B and class 4 lasers and laser systems. Education and training should be provided for operators, maintenance, and service personnel for laser systems containing embedded class 3B and class 4 lasers. The level of training must be commensurate with the level of potential hazard.

10.3.3 Authorized Personnel (Class 1M, 2M, 3R, and 4)

Class 3B and class 4 lasers and laser systems must be operated, maintained, and serviced only by authorized personnel. Lasers and laser systems with enclosed class 3B and class 4 lasers must be maintained and serviced only by authorized personnel if such procedures would permit access to levels that exceed the appropriate MPE.

10.3.4 Alignment Procedures (All Classes Except Class 1)

Alignment of class 2, 3R, 3B, and 4 laser optical systems (mirrors, lenses, beam deflectors, etc.) must be performed in such a manner that the primary beam, or a specular or diffuse reflection of a beam, does not expose the eye to a level above the applicable MPE. Written procedures outlining alignment methods should be approved for class 3B and must be approved for class 4 lasers and laser systems. In the research setting the majority of accidents occur during alignment and other beam manipulation activities.

10.3.4.1 Alignment Procedures for Class 3B and Class 4 Lasers

See Chapter 4 for guidance in addition to the following:

1. Exclude unnecessary personnel from the laser area during alignment.
2. Whenever possible, use low-power visible lasers for path simulation of higher-power visible or invisible lasers.
3. Wear laser protective eyewear during alignment. Use special alignment eyewear when circumstances (e.g., wavelength, power, etc.) permit their use.
4. When aligning invisible (e.g., ultraviolet [UV], infrared [IR]) beams, use beam display devices such as image converter viewers or phosphor cards to locate beams.
5. Perform alignment tasks using high-power lasers at the lowest possible power level.
6. Use a shutter or beam block to block high-power beams at their source except when actually needed during the alignment process.
7. Use a laser-rated beam block to terminate high-power beams downstream of the optics being aligned.
8. Use beam blocks and laser protective barriers in conditions where alignment beams could stray into areas where uninvolved personnel are working.
9. Place beam blocks behind optics (e.g., turning mirrors) to terminate beams that might miss mirrors during alignment.
10. Locate and block all stray reflections before proceeding to the next optical component or section.
11. Be sure all beams and reflections are properly terminated before high-power operation.
12. Post appropriate area warning signs during alignment procedures where lasers are normally class 1 (enclosed).

Alignments should be done only by personnel who have received laser safety training.

10.3.5 Protective Equipment

Enclosure and containment of the laser equipment or beam path is the preferred method of control, since the enclosure isolates or minimizes the hazard. When other control measures do not provide adequate means to prevent access to direct or reflected beams at levels above the MPE, it may be necessary to use personal protective equipment such as eye protection in the form of goggles or spectacles, barriers, windows, clothing and gloves, and other devices that have been specifically selected for suitable protection against laser radiation.

10.3.6 Designation of Areas: Warning Signs and Labels

If a class 3B or class 4 laser is used in an area, then it would be appropriate to segregate that area in some way. The simplest approach is to use the local geography and designate the interior of the room as a laser controlled area. By definition, access will need to be controlled and perhaps restricted to authorized personnel. Such people will usually need to have laser safety training. However, there will need to be a procedure whereby untrained people, such as visitors, can enter.

Permanent designation of an area as a laser controlled area may not be the best solution. It implies that the risk (of eye or skin exposure to the laser beam in excess of the maximum permissible exposure) exists all of the time. In practice this is not going to be the case, even for the most open class 4 product. Sometimes the power to the laser product is switched off. At other times the laser beam will be constrained to particular beam paths. If the risk is very small on most occasions when people enter the area, then they will be unprepared for the occasion when a real risk exists, especially if the entrance door is being used as an alignment target.

10.3.6.1 Signs

Signs can be a very visible form of administrative control. The trouble is that they soon become "wallpaper." Even so, they retain an important aspect of laser-safety management because they are very easy to audit, especially by regulatory authorities.

There are standard and national requirements for safety signs on areas where some laser products are used, as well as requirements under laser safety standards for equipment to be labeled. Signs are only effective if someone first notices them, then reads them, and finally takes some action on the basis of what they have read. Here the term *read* is used in its widest sense since many signs include some form of pictogram.

Where there are regulatory requirements for signs, you have to display them. However, these signs should only be displayed when they are relevant. This can be achieved by either covering signs or turning them over when they are not relevant. Signs falling into this category could include a warning about the laser and the environment, that is, whether it is a laser controlled area; some kind of prohibition sign, such as "authorized persons only"; and perhaps a mandatory sign, "laser safety eyewear must be worn." It may be appropriate to display information about the laser products contained within the room, but generally this is not necessary.

How do you deal with the beam alignment? For many laser applications, this process is carried out by service engineers; in others it is be carried out by the user, especially in research environments. One of the most effective ways of displaying a sign is to get a piece of flip-chart paper and put it on the door with an clear written message such as "Laser Alignment in Progress until 3 pm today [insert today's date]. Please do not enter." This type of sign works for two reasons. First, it is big and new. This means there is a higher probability that someone will notice it. Second, it is time bounded; that is, you have given a finish time. If it looks as if the work is going to go beyond the stated time, then it is better to alter the time than to put up a new notice. The other option is to use a blue and white notice sign as mentioned in the American National Standards Institute's (ANSI) Z136.1 laser standard. If this notice is always displayed on the door and the researcher is known to be in the coffee lounge or on a week's vacation, then the respect for the notice will be lost.

10.3.6.2 Warning Lights

Warning lights can be an improvement on signs, but the message must be clear and the lights must be located where they can be seen. Fitting illuminated panels above doorways is not effective unless the lights flash. There has been a reluctance to fit lights to the side of, or on, doors because they often present a risk of being bumped into by passersby — perhaps even resulting in injuries. However, modern ultra-thin display panels are becoming cost-effective alternatives. They can display warning signs and any other appropriate safety information. They can be made to flash if required and generally can be mounted in any appropriate position.

It is important that any light be labeled so that all who need to know are aware of the meaning. A red flashing light alone does not present any sort of clear message. Does it mean this is a dark room, there is some hazard inside, or "welcome sailors?" In the United States, warning sign wording, color, and layout fall under the guidance of the ANSI Z535 standard.

10.3.6.3 Symbols

Two similar laser symbol designs are accepted for laser signs and labels.

ANSI Z535 Design: This laser hazard symbol is a sunburst pattern consisting of two sets of radial spokes of different lengths and one long spoke radiating from a common center. This is as specified in the ANSI Z535 series of the National Standard Specification for Accident Prevention Signs.

IEC 60825-1 Design: This laser hazard symbol is an equilateral triangle surrounding a sunburst pattern consisting of two sets of radial spokes of different lengths and one spoke radiating from a common center. This is as specified in IEC 60825-1.

10.3.6.3.1 Safety Alert Symbol

This is a symbol that indicates a potential personal safety hazard. It is an equilateral triangle surrounding an exclamation mark and conforms to ANSI Z535.3-1998 Criteria for Safety Symbols. The symbol is to be located to the left of the signal word on the "Danger" or "Caution" signs. It is not used on the "Notice" signs.

10.3.6.4 Signal Word Meanings, as Defined in Section 5 of ANSI

"Danger" indicates an imminently hazardous situation that, if not avoided, will result in death or serious injury. This signal word is to be limited to the most extreme conditions. It must be used with all signs and labels associated with all class 3A lasers and laser systems that exceed the appropriate MPE for irradiance and all class 2M, 3R, and class 4 lasers and laser systems. Some institutions are now listing optical density (OD) on danger signs.

"Caution" indicates a potentially hazardous situation that, if not avoided, may result in minor or moderate injury. It may also be used to alert against unsafe practices. It must be used with all signs and labels associated with class 2 lasers and laser systems and all class 3A lasers and laser systems that do not exceed the appropriate MPE for irradiance.

"Notice" is used to indicate a statement of facility policy, as the message relates directly or indirectly to the safety of personnel or the protection of property. This signal word must not be associated directly with a hazard or hazardous situation and must not be used in place of "Danger" or "Caution." The signal word "Notice" must be used on signs posted outside a temporary laser controlled area. Remember that when a temporary laser controlled area is created, the area outside the temporary area remains class 1, while the area within is either class 3B or class 4, and the appropriate danger warning is also required within the temporary laser controlled area.

10.3.7 Spectators (Class 3B and Class 4)

Protocols for spectators need to be developed to prevent them from being exposed to hazardous laser radiation levels. Items to consider are:

1. Approval has been obtained from the supervisor.
2. The degree of hazard and avoidance procedures have been explained.
3. Appropriate protective measures have been taken.

10.3.8 Laser Radiation Transmitted by Optical Fiber

Laser fiber optics must be considered enclosed systems with the optical cable forming part of the enclosure. If disconnection of a connector results in exposure to a safe level of laser radiation, then no controls are required. Disconnection will yield actual or the potential for laser radiation above a safe level (greater than the MPE). An appropriate laser controlled area needs to be established.

When the connection or disconnection is made by means of a connector other than one within a secured enclosure, such a connector must be disconnected only with the use of a tool. When the connection or disconnection is made within a secured enclosure, no tool is required, but a warning sign appropriate to the class of laser or laser system must be visible when the enclosure is open. Fibers attached to class 3B and class 4 lasers and laser systems must not be disconnected prior to termination of transmission of the beam into the fiber. In this case, if laser radiation above the applicable MPE levels can be made accessible by disconnection of a connector, the connector must be labeled with a tag bearing the words "Hazardous Laser Radiation when Disconnected."

10.3.9 Laser Robotic Installations

It is common to have class 3B and class 4 lasers and laser systems are used in conjunction with robots. In these situations, the robot working envelope (typically 3 to 6 m) should also include the NHZ associated with the laser. In all cases where the beam is focused by a lens associated with the robotic device, appropriate laser-robotic safeguards can be ensured if:

1. The design or control measures in combination provide for a positive beam termination during operation.
2. The beam geometry is limited to only the necessary work task.
3. All workers are located at a distance greater than or equal to the lens-on-laser NHZ value for the laser robotic system. In many robotic use areas this policy is difficult to institute.

10.3.9.1 Location of Equipment Labels

All equipment warning labels must be conspicuously displayed in locations on the equipment where they will best warn onlookers.

10.4 ENGINEERING CONTROLS

In the safety profession it is universally recognized that engineering controls are superior to administrative and procedural controls. The key is that engineering controls remove the human element. Switches that shut off power when covers are removed tend to be more dependable than counting on someone to remember to do the same activity.

Many engineering controls called for in laser safety regulations and standards relate to the controls that are required for laser products. Therefore, at times they present a difficult compliance problem. The R&D atmosphere provides unique challenges to the LSO. The majority of R&D systems are one of a kind, in house, and exempt from the Center of Devices and Radiological Health product safety controls. Users manipulate beams and push the boundaries of output, pulse rate, and so on. Some would argue that for those very reasons a rigid safety protocol is needed, but I believe flexibility and user training will achieve the greatest level of safety. Consider the optical table. The initial laser source maybe a commercial laser or an OEM laser source; the beam can be split, amplified, compressed, stretched, or go through nonlinear optics and change wavelengths, possibility several times over. It is common for that table to have an enclosure around it, maybe to maintain temperature stability, maybe to contain reflections or meet experimental requirements. Sliding a panel back or removing it could expose the user to levels of laser radiation above a level that can cause eye injury. Is that enclosure a protective housing, and should it meet all protective housing requirements? Very few such panels are interlocked. Why? The user may need access to the beams or might need to take diagnostic measurements with beams on.

Many R&D settings do not fit into the classical engineering controls called for in ANSI Z136.1. One size does not always fit all; the R&D LSO needs an open field to tailor the safety program for his or her institution. A tool that lends itself well to the flexibility the LSO needs is the concept of substitution of alternate controls. This option comes from the Z136.1 laser standard and allows the replacement of required controls with an alternative approach reviewed and approved by the LSO. For the LSO, it is a great card to hold. Many times it allows the LSO to obtain a higher level of laser safety compliance from users than if the called-for control was put into practice.

In the R&D setting the best engineering controls are for beam containment. Many others do not apply themselves well to this work environment. The typical laser lab engineering controls are:

1. Entryway controls (interlocks)
2. Illuminated warning signs
3. Beam enclosures
4. Temporary laser controlled areas
5. Laser area warning signs and activation warnings

10.4.1 Access Control (Interlocks)

The purpose of access controls (interlocks) is to protect anyone who enters the laser lab from laser radiation. The user is presented with three options:

Nondefeatable (nonoverride) Area or Entryway Controls: Nondefeatable safety latches, entryway or area interlocks (e.g., electrical switches, pressure sensitive floor mats, infrared, or sonic detectors) must be used to deactivate the laser or reduce the output to levels at or below the applicable MPE in the event of unexpected entry into the laser controlled area.

Defeatable Area or Entryway Controls: Defeatable safety latches, entryway, or area interlocks must be used if nondefeatable area and entryway safety controls limit the intended use of the laser or laser system. If there is no laser radiation hazard at the point of entry, override of the safety controls must be permitted to allow access to authorized personnel provided that they have been adequately trained and provided with adequate personal protective equipment.

Procedural Area Entryway Controls: Where safety latches or interlocks are not feasible or are inappropriate, for example, during medical procedures or surgery, the following apply:

1. All authorized personnel must be adequately trained, and adequate personal protective equipment must be provided upon entry.
2. A door, blocking barrier, screen, curtains, or similar barrier must be used to block, screen, or attenuate the laser radiation at the entryway. The level of laser radiation at the exterior of these devices must not exceed the applicable MPE, nor must personnel experience any exposure above the MPE immediately upon entry.
3. At the entryway there must be a sign or audible signal indicating that the laser is energized and operating at class 4 levels.

Are laser room interlocks required for what are we trying to achieve? No laser lab should be set up such that the act of entering places one in danger. Beams above the MPE (and therefore an eye hazard) should not leave the optical table. One should attempt to keep out unauthorized individuals for their safety and yours. Therefore, card key access, key locks, combination locks, and so on can provide the same level of access control as interlocks keyed to laser power supplies or shutters. Abrupt loss of power to a laser system can cause server damage to the equipment. In addition, only a few commercial laser interlock systems exist, and home-made ones can be very expensive.

10.4.2 Illuminated and Visible Warning Devices

Many forms of illuminated warning devices exist. The oldest style is a single red light. Other options are a lighted laser warning sign that illuminates or flashes when the laser is operating. The light can be electrically interfaced and controlled by the laser power supply so that the light is on and flashing only when the laser is operating. Another possible configuration can be a warning light assembly that may be interfaced to the laser controller to indicate conditions of enabled laser (high voltage on), laser on (beam on), and area clear (no high voltage or beam on).

10.4.3 Beam Enclosures

These should be the laser user's first line of defense, rather than laser protective eyewear. Beam enclosures come in a variety of styles. Beam tubes are one example. So are enclosures around optical tables. The end of the chapter presents a series of beam-control examples.

10.4.4 Temporary Laser Controlled Area

Where removing panels or protective housings, overriding protective housing interlocks, or entering the NHZ becomes necessary (such as for service) and the accessible laser radiation exceeds the applicable MPE, a temporary laser controlled area must be devised for the laser or laser system. Such an area, which by its nature will not have the built-in protective features as defined for a laser controlled area, must provide all safety requirements for all personnel, both within and outside the area.

A notice sign must be posted outside the temporary laser controlled area to warn of the potential hazard.

Following is a list of engineering controls one can find in the ANSI Z136.1 standard. The application of these controls is based on what class laser one is working with. Since in the research setting one is primarily concerned with class 3B and class 4 lasers, we will not consider the lower class lasers for which the majority of these controls do not apply.

10.4.5 Laser Area Warning Signs

The purpose of these signs and devices is to alert those entering the area of the potential laser hazard.

10.5 ENGINEERING CONTROL MEASURES

The following controls are either recommended for class 3B lasers and laser systems or required for class 4 lasers and laser systems. Remember that the engineering and administrative development of control measures should be a team effort between the LSO and users, particularly in an active research setting:

1. Protective housings
2. Without protective housing
3. Interlocks on removable protective housings
4. Service access panel control
5. Viewing windows, display screens, and collecting optics
6. Collecting optics
7. Totally open beam path
8. Limited open beam path
9. Remote interlock connector
10. Beam stop or attenuator

11. Activation warning system
12. Indoor laser controlled area
13. Class 3B indoor laser controlled area
14. Class 4 laser controlled area
15. Laser outdoor controls
16. Laser in navigable airspace
17. Temporary laser controlled area
18. Controlled operation
19. Equipment labels
20. Laser area warning signs and activation warnings

10.5.1 Protective Housings

A protective housing must be provided for all classes of lasers and laser systems. The protective housing may require interlocks and labels. Special safety procedures may be required when protective housings are removed, for example, for alignment, such as a temporary control area.

10.5.2 Operating a Laser without Protective Housing

While protective housings are required for all classes of lasers, it is common to operate R&D lasers in violation of this requirement. You will find other circumstances in addition to R&D such as during the servicing of lasers, in which operation of lasers and laser systems without a protective housing is common. In such cases the LSO must effect a hazard analysis and ensure that control measures are instituted.

10.5.2.1 Walk-In Protective Housing (Embedded Class 3B and Class 4)

Class 1 lasers and laser systems that contain embedded class 3B or class 4 lasers with protective housings of sufficient size to allow personnel within the working space (walk-in protective housings) must be provided with an area warning system (floor mats, IR sensors, etc.) that is activated upon entry by personnel into the protective housing. The sensors must be designed to interlock with the laser power supply or laser shutter so as to prevent access to laser radiation above the applicable MPE. Only authorized personnel can be provided with the means to override the sensors for alignment or testing procedures if beam access is required for beam diagnostic purposes.

10.5.3 Interlocks on Removable Protective Housings (All Classes)

Protective housings that enclose class 3B or class 4 lasers and laser systems must be provided with an interlock system that is activated when the protective housing is opened or removed during operation and maintenance. These interlocks may be electrically or mechanically interfaced to a shutter that interrupts the beam when the protective housing is opened or removed.

10.5.4 Service Access Panels (All Classes)

Portions of the protective housing that are only intended to be removed from the laser or laser system by service personnel, who then permit direct access to laser radiation associated with a class 3B or class 4 laser or laser system, must either: (a) be interlocked (fail-safe interlock not required), or (b) require a tool for removal and have an appropriate warning label. The goal is that the housing should not be easy to open like a bathroom cabinet. Rather, one needs to demonstrate some forethought to open the housing, such as using a tool.

10.5.4.1 Key Control

A class 3B or class 4 laser or laser system must be provided with a master switch. This master switch affects beam termination and system shutoff and must be operated by a key or by a coded access (such as a computer code).

As an alternative, the master switch can be designed to allow system activation using a momentary switch action (or alternative) that initiates system operation with the option that the key (or alternative) can be removed after operation commences. In this mode, if the system ceases to operate, the key switch (or alternative) must again be used to restart the laser or laser system. A single master switch on a main control unit should be acceptable for multiple laser installations where the operational controls have been integrated. This allows the designation of a circuit breaker to be compliant with this control.

10.5.5 Viewing Windows, Display Screens, and Collecting Optics

This includes lenses, telescopes, microscopes, endoscopes, and so on.What one is looking for here is assurance that levels of laser radiation above MPE are not transmitted through these items. This is why it is so important for the user to know what wavelengths are in use and their transmission properties through various materials.

10.5.6 Limited Open Beam Path (Class 3B and Class 4)

In applications of class 3B or a laser hazard analysis by the LSO is required.

10.5.6.1 Class 1 Conditions

Frequently the hazard analysis defines an extremely limited NHZ, and procedural controls can provide adequate protection. Class 1 conditions must be considered as fulfilled for those limited open beam path lasers or laser systems where analysis confirms that the accessible levels during operation are at or below applicable MPE levels.

10.5.6.2 Enclosed Beam Path (All Classes)

In applications of lasers or laser systems where the entire beam path is enclosed and the enclosure fulfills all requirements of a protective housing (i.e., limits the laser radiation exposure at or below the applicable MPE), the requirements of class 1 are fulfilled and no further controls are required. When the protective housing requirements are temporarily relaxed, such as during service, the LSO must carry out the appropriate controls.

10.5.7 Remote Interlock Connector (Class 3B and Class 4)

This is an equipment manufacturing design element found in many class 3B and Class 4 lasers, but not all. A class 3B laser or laser system should and class 4 lasers and laser systems must be provided with a remote interlock connector. The interlock connector facilitates electrical connections to an emergency master disconnector interlock or to a room, entryway, floor, or area interlock, as may be required for a class 4 controlled area.

10.5.8 Beam Stop or Attenuator (Class 3B or Class 4)

These devices play a major role in containing reflections. Labeling of beam stops is a useful tool used by many laser users.

10.5.9 Laser Area Warning Signs and Activation Warnings

10.5.9.1 Class 3B and Class 4 Laser Areas

An area that contains a class 3B or class 4 laser or laser system must be posted with a laser warning sign. The location should be easily visible.

10.5.9.2 Temporary Laser Controlled Area (Class 3B and Class 4)

The exterior boundary of a temporary laser controlled area shall be posted with a notice sign.

10.5.9.3 Laser Warning Sign Purpose (Class 3B and Class 4)

The purpose of a laser area warning sign is to convey a rapid, visual hazard-alerting message:

1. Warns of the presence of a laser hazard in the area
2. Indicates specific policy in effect relative to laser controls

3. Indicates the severity of the hazard (e.g., class of laser, NHZ extent, etc.)
4. Instructs appropriate action to take to reduce the hazard (eyewear requirements, etc.)

10.5.9.3.1 Warning Signs for Nonbeam Hazards (Class 3B and Class 4)

Warning signs for nonbeam hazards (e.g., high voltage, cryogenics, and compressed gases) must be posted when the hazards are possible as specified in such documents as ANSI 535.2 (see Multiple Hazard Labeling Requirements) and other standards applicable to the specific hazard.

10.5.9.3.2 Audible Warning Devices (Class 3B and Class 4)

While this control is listed in the ANSI Z136 standard, it is rarely used, and communication within the laser use area is more important and useful.

10.5.9.3.3 Visible Warning Devices (Class 3B and Class 4)

Several types of visible warning lights exist, and better ones will be developed in the near future. The majority are found outside the laser use area to warn those who may wish to enter about the potential laser hazard within. Some settings have a similar sign on the inside so all those in the laser use area are aware of the laser system's status. One of the oldest types of warning light is a single red light or lighted laser warning sign that flashes when the laser is operating. Internal signs need to be readably visible through laser protective eyewear and viewable within the area. Another option is a digital scroll sign or LED display panel. The light can be electrically interfaced and controlled by the laser power supply so that the light is on and flashing only when the laser is operating. Another possible configuration is a warning light assembly that may be interfaced to the laser controller to indicate conditions of enabled laser (high voltage on), laser on (beam on), and area clear (no high voltage or beam on). In this case, a green light indicates when the laser is not operational (high voltage off), and a yellow light indicates when the laser is powered up (high voltage applied, but no laser emission). An additional (flashing optional) red light is activated when the laser is operating. The LSO and area supervisor need to be aware of alternative control measures for the hearing and visually impaired.

10.5.9.3.4 Indoor Laser Controlled Area (Class 3B and Class 4)

A laser hazard analysis must be conducted by the LSO. If the analysis determines that the classification associated with the maximum level of accessible radiation is class 3B or class 4, a laser controlled area must be established and adequate control measures instituted.

10.5.9.3.5 Class 3B Indoor Laser Controlled Area (Class 3B)

The class 3B laser controlled area must:

1. Be controlled to permit lasers and laser systems to be operated only by personnel who have been trained in the operation of the laser, laser system, and laser safety
2. Be posted with the appropriate warning sign(s) and, if deemed necessary by the LSO, posted within the laser-controlled area
3. Be operated in a manner such that the path is well defined
4. Allow for the beam to be well defined and controlled if the beam extends beyond an indoor area and projects into a controlled airspace, particularly under adverse atmospheric conditions (i.e., rain, fog, snow, etc.)

In addition, a class 3B controlled area should:

5. Be under the direct supervision of an individual knowledgeable in laser safety
6. Be located so that access to the area by spectators is limited and requires approval
7. Have any potentially hazardous beam terminated in a beamstop of an appropriate material
8. Have only diffusely reflecting materials in or near the beam path, where feasible
9. Provide personnel within the laser controlled area with the appropriate eye protection
10. Have the laser secured such that the exposed beam path is above or below the eye level of a person in any standing or seated position, except as required for medical use
11. Have all windows, doorways, open portals, and so on from an indoor facility be either covered or restricted in such a manner as to reduce the transmitted laser radiation to levels at or below the applicable ocular MPE
12. Require storage or disabling (for example, removal of the key) of the laser or laser system when not in use to prevent unauthorized use

10.5.9.3.6 Class 4 Laser Controlled Area (Class 4)

All class 4 area or entryway safety controls must be designed to allow both rapid egress by laser personnel at all times and admittance to the laser controlled area under emergency conditions. All personnel who require entry into a laser controlled area must be appropriately trained, provided with appropriate protective equipment, and follow all applicable administrative and procedural controls.

For emergency conditions there must be a clearly marked “Emergency Stop” or other appropriately marked device appropriate for the intended purpose (remote-controlled connector or equivalent device) available for deactivating the laser or reducing the output to levels at or below the applicable MPE. R&D posted shut-door procedures are a better option than E stop.

10.5.10 Laser Outdoor Controls (All Classes)

All classes of lasers and laser systems used outdoors must meet the following requirements:

1. The LSO must conduct an analysis to establish the NHZ if it is not provided as part of the documentation furnished by the manufacturer. If visible lasers are used at night, the LSO will conduct an analysis to determine if the laser beams visually interfere with critical tasks. For operation of visible lasers at night near airports, refer to Federal Aviation Administration (FAA) Order 7400.2 and ANSI Z136.6.
2. The NHZ must be clearly posted with laser warning signs and demarcated and identified as the laser hazard area. All personnel authorized to enter the NHZ must be appropriately trained. Only personnel who have been authorized may operate a laser or laser system. Appropriate combinations of physical barriers, screening, and personal protective equipment must be provided and used by those personnel.
3. Appropriate administrative controls must be used if personnel are permitted within the NHZ. Directing the laser beam toward automobiles, aircraft, or other manned structures or vehicles must be prohibited within the NHZ unless adequate training and protective equipment is provided and used by all personnel within these targets or as authorized by the LSO and permitted by FAA Order 7400.2. In such authorized cases, it is essential that adequate training and protective equipment be provided and used by all personnel within these areas. The exposed laser beam path must not be maintained at or near personnel eye level without specific authorization of the LSO.
4. The beam path must be confined and terminated wherever possible When the laser is not being used, it must be disabled in a manner that prevents unauthorized use.
5. The operation of class 4 lasers and laser systems during rain, snow, or fog or in a dusty atmosphere may produce hazardous scattering near the beam. In such conditions, the LSO must evaluate the need for, and specify the use of, appropriate personnel protective equipment.

10.5.11 Use of Lasers in Navigable Airspace

The FAA is responsible for regulating the use and efficient utilization of navigable airspace to ensure the safety of aircraft and the protection of people and property on the ground. Laser experiments and programs that involve the use of lasers or laser systems in navigable airspace should be coordinated with the FAA. Refer to FAA Order 7400.2 and ANSI Z136.6.

10.5.12 Temporary Laser Controlled Area (All Classes)

When removal of panels or protective housings, overriding of protective housing interlocks, or entry into the NHZ becomes necessary (such as for service) and the accessible laser radiation exceeds the applicable MPE, a temporary laser controlled area must be devised for the laser or laser system. Such an area, which by its nature will not have the built-in protective features as defined for a laser controlled area, must provide all safety requirements for all personnel, both within and outside the area. A notice sign must be posted outside the temporary laser controlled area to warn of the potential hazards.

10.5.13 Controlled Operation (Class 4)

Whenever appropriate and possible, class 4 lasers and laser systems should be controlled and monitored at a position as distant as possible from the emission portal of the laser or laser system.

10.5.14 Equipment Labels (All Classes)

The majority of laser equipment found in the United States is labeled in accordance with the Federal Laser Product Performance Standard (FLPPS).

10.5.14.1 Logotype Warning Equipment Label (All Classes Except Class 1)

All lasers and laser systems (except class 1) must have appropriate warning labels with the laser sunburst logotype symbol and the appropriate cautionary statement. The label must be affixed to a conspicuous place on the laser housing or control panel. Such labels should be placed on both the housing and the control panel if these are separated by more than 2 m.

10.5.14.2 Protective Housing Equipment Label (All Classes)

An advisory protective housing label that indicates the relative hazard of laser radiation contained within the housing must be placed on all removable protective housings that have no safety interlock and that can be removed or displaced during maintenance or service and thereby allow access to laser radiation in excess of the applicable MPE. The laser sunburst logotype symbol is not required on such advisory labels.

10.5.14.2.1 Long-Distance Beam Conduit Label (All Classes Except Class 1)

The LSO must post advisory protective housing labeling on long-distance (>3 m) beam conduits that contain beams operating above class 1 levels. Such labeling shall be placed on the outside of the conduit at appropriate intervals (approximately 3 m) to warn of the relative hazards of laser radiation contained within the conduit. The laser sunburst logotype symbol is not required on such advisory protective housing labels. (For practical control measures examples see Chapter 14.)

11 U.S. and European Regulations and Standards

Trust us we are from the government and know what's best.

11.1 INTRODUCTION

In the United States laser regulations fall into two groups: those directed to laser manufacturers (which sometimes can also be laser users) and those directed to laser users. In addition a number of nongovernmental user guideline standards exist. Some of these standards, such as the American National Standards Institute (ANSI) Z136.1 Safe Use of Lasers, may have a greater impact than direct governmental regulations.

11.2 USER REGULATIONS

11.2.1 The Occupational and Safety Health Administration

The Occupational and Safety Health Administration (OSHA) is a federal government agency. It is empowered by the Occupational and Safety Health Act. The Occupational and Safety Health Act's mandate is to ensure that worker and workplace safety is addressed. The act provides that workers must work in a place of employment free from recognized hazards to safety and health, such as exposure to toxic chemicals, excessive noise levels, mechanical dangers, heat or cold stress, and unsanitary conditions. OSHA has limited guidance directed to laser radiation. Some states have OSHA approved state plans and have adopted their own standards and enforcement policies. In these states federal OSHA relinquishes its authority to the state governmental body; for example, California has Cal OSHA.

The General Duty Clause Section 5(a) (1) requires employers to provide employment free from recognized hazards that could cause serious physical harm. Using the General Duty Clause, OSHA often relies on published voluntary guidelines to determine what constitutes an environment free from recognized hazards. For lasers, the industry standard OSHA holds an employer to is compliance with the ANSI Z136.1 Safe Use of Lasers standard. OSHA relies on its General Duty

Clause to make up for its lack of specific laser regulations covering the breadth of laser safety requirements. While OSHA relies on the General Duty Clause for enforcement, it has published several technical documents to aid the laser user as well as the OSHA inspector:

Guidelines for Laser Safety and Hazard Assessment: OSHA STD 01-05-001 - PUB 8-1.7 (August 5, 1991). This provides guidelines to federal OSHA and State compliance officers, consultants, and employees for the assessment of laser safety. It is a tutorial on lasers.

Guidelines for Robotics Safety: OSHA STD 01-12-002 - PUB 8-1.3 (September 21, 1987). This document identifies controls that can be used on robotic laser systems.

Technical Directorate: TED 1.15, Section II, Chapter 6. This deals with laser hazards. It is a somewhat trimmed-down version of publication 8.1.7.

Technical Manual: (TED 1-0.15A), Section III, Chapter 6, Appendix III:6-4 (January 20, 1999). This manual provides a few examples of warning signs that can be used to alert workers of potential laser hazards.

OSHA has two general regulatory areas on personnel protective equipment that, while not specific to lasers, can be applied to laser protective eyewear. These are found in the general industry section, 1910.132, General Requirements for Personal Protective Equipment and 1910.133, Eye and Face Protection. The construction industry section, 1926.54, Nonionizing Radiation, contains a number of general laser guidelines:

1. Only qualified and trained employees shall be assigned to install, adjust, and operate laser equipment.
2. Proof of qualification of the laser equipment operator shall be available and in the possession of the operator at all times.
3. Employees, when working in areas in which a potential exposure to direct or reflected laser light greater than 0.005 watts (5 milliwatt) exists, shall be provided with antilaser eye protection devices as specified in Subpart E of this part.
4. Areas in which lasers are used shall be posted with standard laser warning placards.
5. Beam shutters or caps shall be utilized, or the laser turned off, when laser transmission is not actually required. When the laser is left unattended for a substantial period of time, such as during lunch hour, overnight, or at change of shifts, the laser shall be turned off.
6. Only mechanical or electronic means shall be used as a detector for guiding the internal alignment of the laser.
7. The laser beam shall not be directed at employees.
8. When it is raining or snowing, or when there is dust or fog in the air, the operation of laser systems shall be prohibited where practicable;

in any event, employees shall be kept out of range of the area of source and target during such weather conditions.

9. Laser equipment shall bear a label to indicate maximum output.
10. Employees shall not be exposed to light intensities above:
    - Direct staring: 1 micro-watt per square centimeter;
    - Incidental observing: 1 milliwatt per square centimeter;
    - Diffused reflected light: 2 1/2 watts per square centimeter.
11. Laser units in operation should be set up above the heads of the employees, when possible.

OSHA also has a number of laser related interpretations and position letters. A sampling of topics represented is:

1. Issuance of and paying for personal protective equipment (PPE). OSHA (December 9, 1996), one page. This is a response to a letter concerning the issuance of PPE and who should pay for this equipment.
2. Hazards of smoke generated from surgical procedures. OSHA (April 18, 1996), one page. This is a letter responding to a question about requirements addressing hazards of smoke generated from surgical procedures.
3. Clarifications of the terms *direct staring* and *incidental observing*. OSHA (June 13, 1994), one page. This response concerns clarification of the standard addressing the use of lasers [29 CFR 1926.54(j)].
4. Class 1 lasers. OSHA (July 12, 1993), one page. This letter covers regulations addressing class 1 lasers.
5. "De minimis violation" status, which is accorded to power lasers used in the construction industry. OSHA (October 21, 1992), one page. This letter deals with class 1 lasers resulting in de minimis violation.
6. Necessary training for laser equipment operators. OSHA (September 21, 1992), one page. This letter addresses training requirements for the use of level 1 lasers.
7. Laser standards applicable to a surgical laser program. OSHA (June 3, 1987), one page. When this letter was written, OSHA did not have a standard regulating employee use of lasers.
8. Supermarket laser scanning equipment. OSHA (April 29, 1985), one page. No specific standards regulate this equipment. This letter directs the question to state departments of labor and the Department of Public Health administrator.
9. Phototropic lenses in safety eye wear. OSHA (April 16, 1976), one page. This letter addresses the requirements of proper PPE.

It is clear that state and federal OSHA recognize lasers as a workplace hazard that needs to be addressed. The default position these agencies have taken is to seek compliance with the ANSI Z136.1 standard to meet their goal of worker safety rather than generate more regulations. One reason for this is the ANSI policy of routine updates of the standard based on the latest biological and control measures data. While being on an approximately 5-year cycle, it is still faster than many regulatory bodies move. All of this increases the importance of the ANSI standard to users within the United States.

11.2.2 Federal Aviation Administration (FAA)

The FAA is responsible for the safety of U.S. navigable airspace (NAS). With respect to the use of lasers outdoors, for years this responsibility lay primarily with the Center for Devices and Radiological Health (CDRH). Starting in 2000, the FAA increased its role and presence on the regulatory scene with FAA orders 7400.2E "Procedures for Handling Airspace Matters," Chapter 28 "Outdoor Laser Operations."

The FAA procedure is for regional offices to conduct an aeronautical review of all laser operations to be performed in their NAS to ensure that these types of operations will not have a detrimental effect on aircraft operations. The official guidance the FAA has adopted is, "For full consideration to be given to national defense requirements, commercial uses, and general aviation operations that have the public right of 'freedom of transit' through the NAS."

Accordingly, while an effort must be made to negotiate equitable solutions to conflicts over the use for nonaviation purposes, preservation of the NAS for aviation must receive primary emphasis (Figure 11.1).

The FAA has defined a number of zones:

Laser free zone (LFZ): Airspace in the immediate proximity of the airport, up to and including 2000 feet above ground level (AGL), extending 2 nautical miles (NM) in all directions measured from the runway centerline. Additionally, the LFZ includes a 3-NM extension, 2500 feet to each side of the extended runway centerline, up to 2000 feet AGL of each usable runway surface. The level of laser light is restricted to a level that should not cause any visual disruption.

Critical flight zone (CFZ): Airspace within a 10-NM radius of the airport reference point, up to and including 10,000 feet AGL, where a level of laser light is restricted to avoid flash blindness or afterimage effects.

Sensitive flight zone (SFZ): Airspace outside the CFZs that authorities (e.g., FAA, local departments of aviation, military) have identified that must be protected from the potential effects of laser emissions.

Normal flight zone (NFZ): Airspace not defined by the LFZs, CFZs, or SFZs.

11.2.2.1 Flight Zone Exposure Distance

This is the maximum distance from the laser system beyond which the laser beam's irradiance level does not exceed a specific level:

1. Laser free zone: 50nW/cm^2
2. Critical zone: 5μW/cm^2
3. Sensitive zone: 100μW/cm^2

Civilian users of lasers crossing into open space should obtain a copy of ANSI Z136.6, Safe Use of Lasers Outdoors. It provides a wide variety of guidance for a host of outdoor applications.

11.3 STATE REGULATIONS

A number of states have some sort of laser regulations; this excludes the occasional inspection of laser light shows. These regulations are either at the state government or local municipality level. Local regulations tend to deal with the use or purchase of laser pointers. A number of state government agencies have either laser regulations or an agency that has statutory authority to regulate laser use, but no regulations have been created. States with rules require registration and compliance with laser user policies based on the ANSI Z136.1 standard. Arizona and Massachusetts are examples of states with compressive programs.

Even states without laser regulations for users may have specific regulations for medical laser applications, for example, laser hair removal. For laser hair removal, as many states restrict laser use to physicians as have no restrictions. Here is a sampling of U.S. states that have regulatory programs:

Alaska: Radiological Health Program Section of State Laboratories Department of Health and Social Services, Title 18 of the Alaska Annotated Code Part 85, Art. 7, Sect. 670-730 (October 1971 and April 1973)

Arizona: Radiation Regulation Agency regulations. Article 14, Rules for Control of NIR, Sect. R12-1-1421 to 1444

Florida: Department of Health, Bureau of Radiation Control regulations. Extensive regulations in Chapter 10D-89 of the Florida code

Georgia: Office of Regulatory Service Department of Human Resources regulations. Registration requirements in Chapter 270-5-27, Georgia Code (9/1/71)

Illinois: Division of Electronic Products, Department of Nuclear Safety Registration regulations in Laser Systems Act of 1997 (effective July 25, 1997)

Massachusetts: Massachusetts Radiation Control Program regulations. Registration and control regulations (ANSI Z136 based) (effective May 7, 1997)

New York: Department of Labor Radiological Health Unit regulations. In Industrial Code Rule 50 of Title 12 (cited 12 NYCRR Part 50; amended March 2, 1994)

Texas: Bureau of Radiation Control, Department of Health Division of Licensing, Registration and Standards.Texas Regulations for the Control of Laser Hazards (TRCLRH) Part 50, 60, and 70. -25 Texas Administrative Code (TAC) 289.301

11.4 USER STANDARDS

A number of standards exist for laser user guidance; the most important of these are the ANSI Z136.1 standards. These standards provide guidance for the safe use of lasers and laser systems between 180 and 1 mm. The ANSI Z136 committee is made up of volunteers from industry, government, and public sectors. An active process is in effect to prevent the committee from having too many representatives from one sector or another, that is, military versus academia. ANSI Z136.1 for safe use of lasers is a horizontal laser safety standard; it is designed to give the laser user guidance on obtaining laser safety. ANSI Z136.1 is the most cited laser safety guidance in the United States. It is the basis of laser safety for the Department of Energy, Department of Labor, U.S. military and state regulatory agencies, industrial workplaces, and most universities. In short, laser safety cannot be accomplished in the United States without some understanding of the ANSI Z136.1 standard or part of its series.

Some of the most important features of the ANSI Z136.1 standards are:

1. The establishment of a laser safety officer (LSO) role and responsibilities
2. Recognition that management is responsible to provide resources to the LSO
3. Guidance on laser safety training for the LSO, users, and ancillary personnel
4. The concept of alternate controls in place of controls called for in the standard when the LSO and use conditions deem greater flexibility
5. Control measures for class 3B and class 4 lasers or laser systems, which are listed per laser class and divided between the following:
   a. Engineering controls, which in many cases are compatible with CDRH requirements
   b. Administrative controls, which fall into many sections of laser safety program management
6. Updates on safe exposure levels termed maximum permissible exposure (MPE)
7. Official design of laser warning signs
8. Explanation of laser hazard classifications

Over the years as laser applications have increased, a number of vertical standards have been developed. Each of these to varying degrees refers back to the Z136.1 document but addresses important considerations for its respective applications. A list of those standards includes:

ANSI Z136.2: American National Standard for Safe Use of Optical Fiber Communications Systems Utilizing Laser Diodes and LED Sources

ANSI Z136.3: American National Standard for Safe Use of Lasers in Health Care Facilities

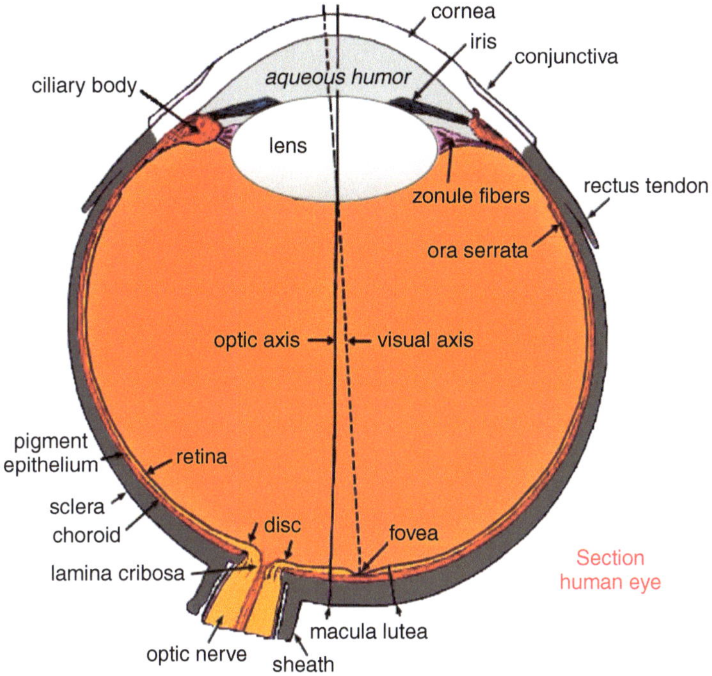

COLOR FIGURE 2.1 Human eye components.

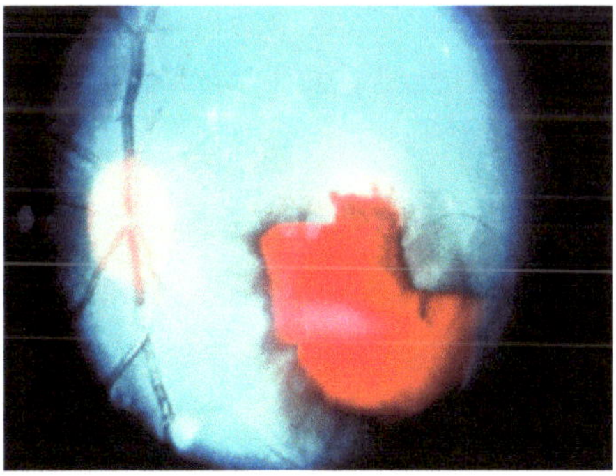

COLOR FIGURE 2.5 Blood in vitreous.

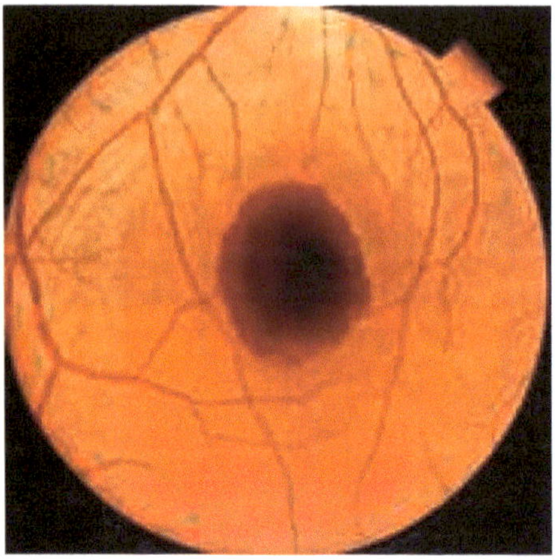

COLOR FIGURE 2.6 Pooled blood.

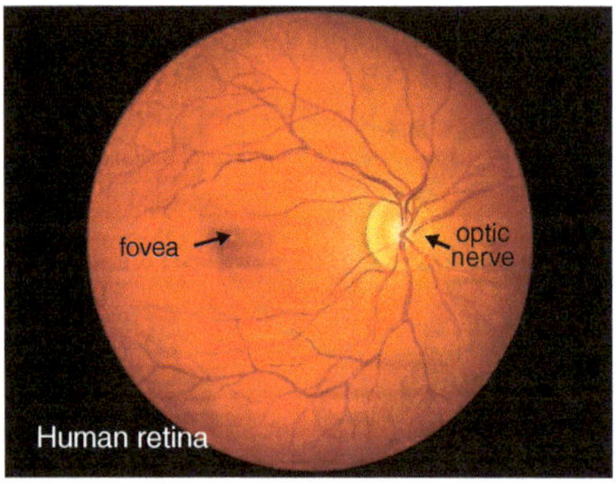

COLOR FIGURE 2.7 Retinal injuries.

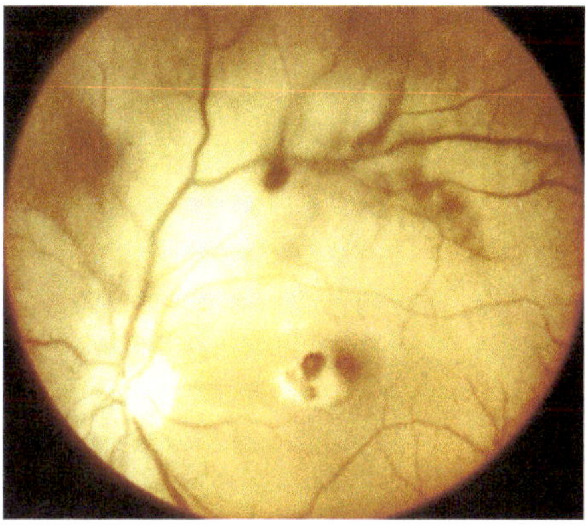

COLOR FIGURE 2.8 Retinal injuries.

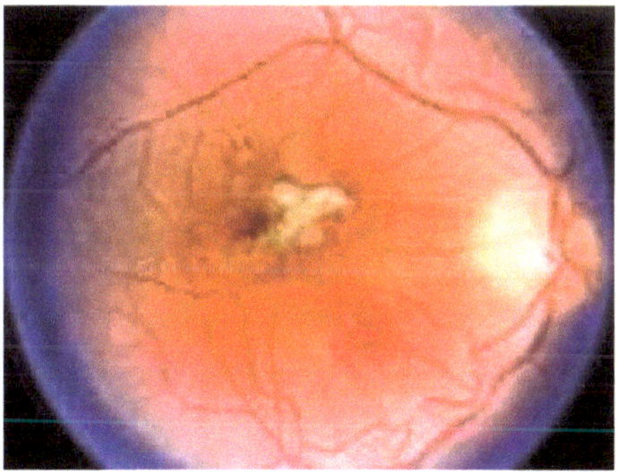

COLOR FIGURE 2.9 Severe fovea and macula damage from NIR exposure.

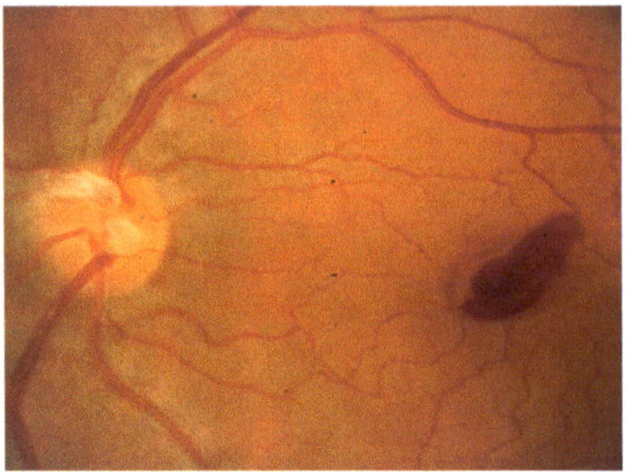

COLOR FIGURE 2.10 Retinal injuries.

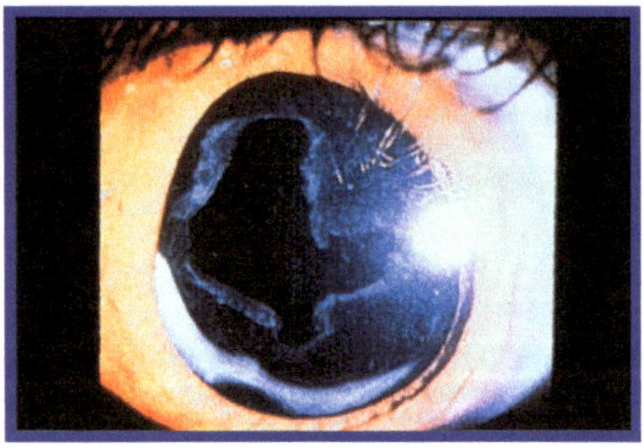

COLOR FIGURE 2.11 Corneal burn.

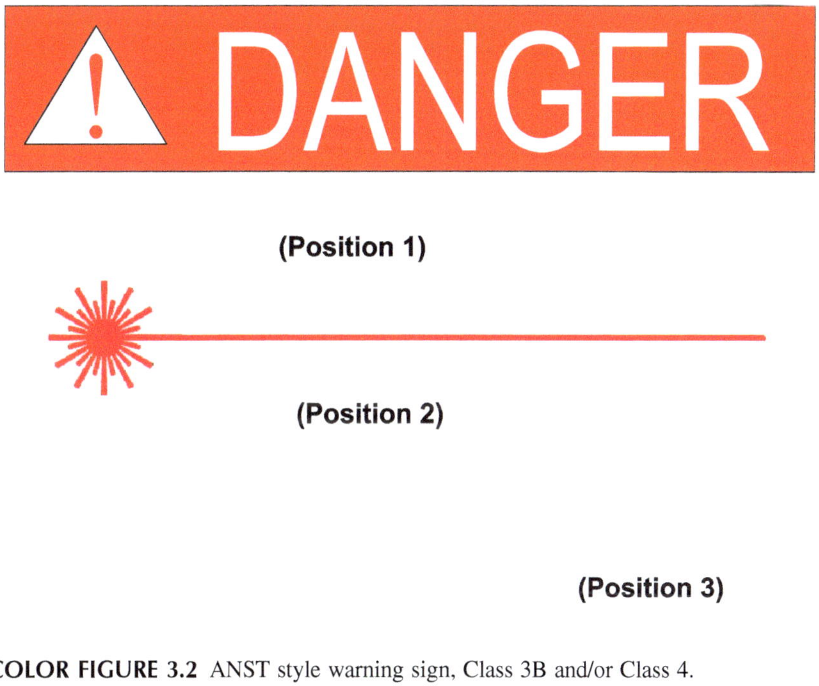

COLOR FIGURE 3.2 ANST style warning sign, Class 3B and/or Class 4.

COLOR TABLE 3.1 Specific Instructions for the Formatting of Laser Area Warning Signs

Safety Alert Symbol & Signal Word	Position 3 Laser Class *right justify, 18-pt*	Position 1 (bold words are ANSI format) *center justify, 18-pt; also see Note below*	Position 2 *left justify, 18-pt*	Reason for use, Comments	Safety symbol
Colors to be used	Use **bold black Helvetica upper & lower case**; background color is white or yellow as shown				
⚠ DANGER Red & white	**Class 4 laser**	**Invisible and Visible Laser Radiation — Avoid Eye or Skin Exposure to Direct or Scattered Radiation**	Information about access, OD requirements, OSP #, etc.; list lasers, OD, wavelengths (or pulse, PRF, output, etc.)	When highest class laser is Class 4	Red
⚠ DANGER Red & white	**Class 3b laser**	**Invisible and Visible Laser Radiation — Avoid Direct Exposure to Beam**	"	When highest class laser is Class 3b	Red
⚠ DANGER Red & white	**Class 3a laser**	**Invisible and Visible Laser Radiation — Avoid Direct Eye Exposure**	"	When highest class laser is Class 3a and above MPE	Red
⚠ CAUTION Yellow & black	**Class 3a laser** *(this sign is seldom used - see comments)*	**Visible Laser Light — Do Not Stare into Beam or View Directly with Optical Instruments**	"	When highest class laser is *visible* Class 3a *and* below MPE for 0.25 s *Note: yellow background*	Black
⚠ CAUTION Yellow & black	**Class 2 laser** *(Class 2a has been eliminated)*	**Visible Laser Light — Do Not Stare into Beam**	"	When highest class laser is Class 2 (all Class 2 lasers are visible) *Note: yellow background*	Black
CAUTION Yellow & black; no alert symbol	N/A	Describe the nature of hazards other than personal injury	Information about potential property damage	For advice on property damage only *Note: yellow background*	Black symbols representative of hazard
⚠ WARNING Orange & white	N/A	Describe the nature of hazards (e.g. controls for servicing ion lasers; unattended laser operation, etc.); provide access instructions or direct to other detailed instructions	Contact information in case of emergency	For advice on potential personal injury if not avoided	Black
NOTICE Blue & white; no alert symbol	Date and name of person posting if a temporary notice.	Describe the nature of the policy or temporary condition (this may supplement DANGER or CAUTION signs but *not* replace them)	Actions to be taken, if necessary, such as before entering	Policy statement; interlocks or status panel not needed; posting outside a temporary laser controlled area; etc.	Black (use symbol if laser related)

COLOR FIGURE 5.1 (a) Laser danger sign as of 2000; (b) notice sign: used for repair or alignment notification.

COLOR FIGURE 8.1 Laser eyewear, Glendale laser products.

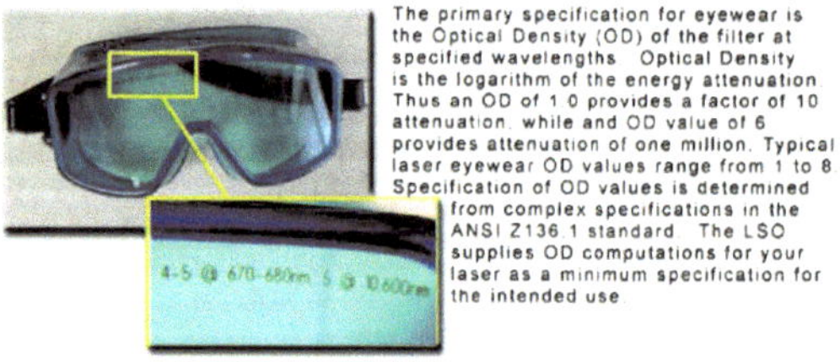

COLOR FIGURE 8.2 Labeling.

One burst of energy from a CO_2 laser is enough to ignite polyvinyl chloride tubing and turn it into a deadly blowtorch. The tubing should not be used in the airways of patients undergoing laser surgery of the upper or lower respiratory tract. Courtesy of Dr. Daniel E. Supkis Jr. and the M.D. Anderson Cancer Center.

COLOR FIGURE 13.1 Endotrachel tube on fire.

COLOR FIGURE 14.1 Homemade beam block.

COLOR FIGURE 14.4 View from inside enclosed table.

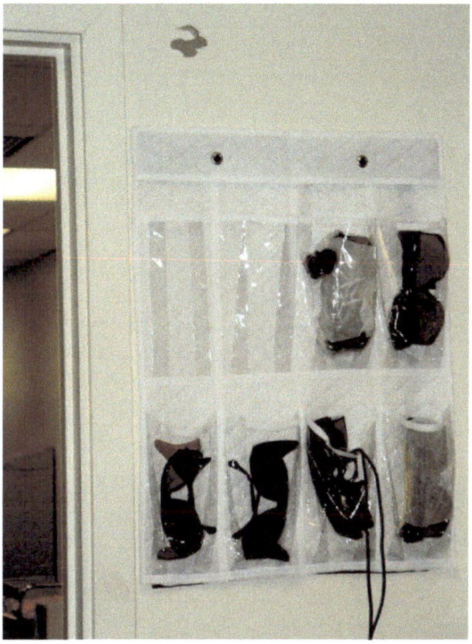

COLOR FIGURE 14.15 Similar eyewear holder.

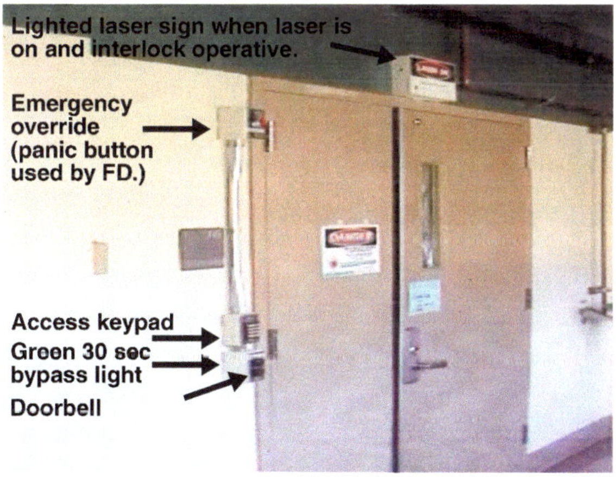

COLOR FIGURE 14.16 Sign too high; binder on door.

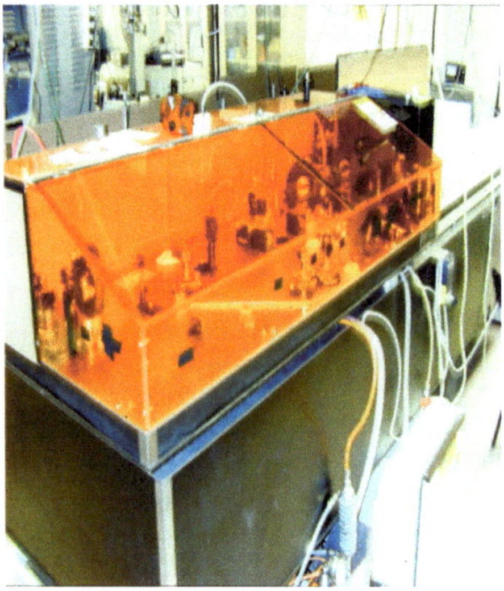

COLOR FIGURE 14.19 Acrylic enclosure.

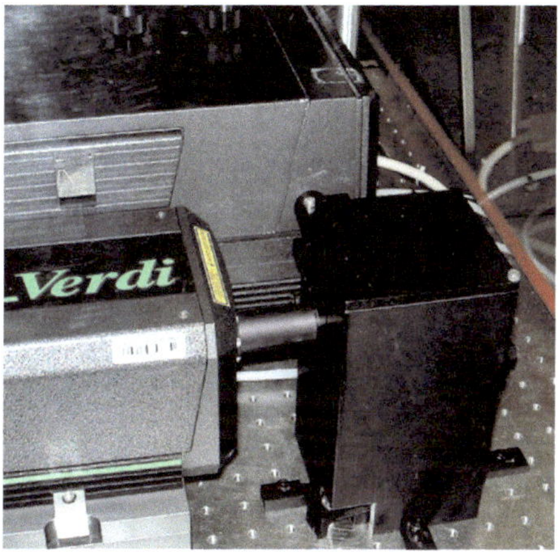

COLOR FIGURE 14.20 Optics and beam path enclosed.

COLOR FIGURE 14.31 Dye stain

COLOR FIGURE 14.32 Secondary container for dye pumps.

COLOR FIGURE 14.39 External shutter.

ANSI Z136.4: American National Standard for Laser Safety Measurements and Instrumentation
ANSI Z136.5: American National Standard for Safe Use of Lasers in Educational Institutions
ANSI Z136.6: American National Standard for Safe Use of Lasers Outdoors
ANSI Z136.7: American National Standard for Certification and Testing of Laser Eyewear and Barriers

In addition, a number of other laser user standards exist, separate from but related to the Z136 series:

1. National Fire Protection Association code #115 preventing laser fires
2. ANSI/American Welding Society C.7.2 Recommended Practices for Laser Beam Welding, Cutting, and Drilling
3. ANSI B11.21 Machine Tools-Machine Tools Using Lasers for Processing Materials-Safety Requirement for Design, Construction, Care and Use
4. SEMI, the semiconductor industry laser use standard

11.5 MANUFACTURER REGULATIONS

A division of the U.S. Food and Drug Administration (FDA) is the Center for Devices and Radiological Health (CDRH). The CDRH is responsible for regulating firms that manufacture, repackage, relabel, and import medical devices sold in the United States. In addition, the CDRH regulates radiation-emitting electronic products (medical and nonmedical) such as lasers, x-ray systems, ultrasound equipment, microwave ovens, and color televisions. It is easy to comprehend the vast number of products the CDRH evaluates and monitors, because the CDRH also monitors devices throughout the product lifecycle, including a nationwide postmarket surveillance system. It ensures that radiation-emitting products and laser products meet radiation safety standards.

Manufacturers and distributors of products meeting the definition of "electronic product radiation" in Section 531 of the Federal Food, Drug, and Cosmetic (FD&C) Act may be subject to certain provisions of the act including the retention of records and submission of product reports to the FDA, specifically to the CRDH. The FDA's requirements for these products, record keeping, and reporting are included in the final regulations contained in Title 21 Code of Federal Regulations Parts 1000-1299 (21 CFR 1000- 1299). According to section 531 of the FD&C Act:

- (1) The term "electronic product radiation" means
  - (A) any ionizing or non-ionizing electromagnetic or particulate radiation, or

  - (B) any sonic, infrasonic, or ultrasonic wave emitted from an electronic product as the result of the operation of an electronic circuit in such product;
- (2) The term "electronic product" means
  - (A) any manufactured or assembled product which, when in operation,
    - (i) contains or acts as part of an electronic circuit and
    - (ii) emits (or in the absence of effective shielding or other controls would emit) electronic product radiation, or
  - (B) any manufactured or assembled article which is intended for use as a component, part, or accessory of a product described in clause (A) and which when in operation emits (or in the absence of effective shielding or other controls would emit) such radiation;
- (3) The term "manufacturer" means any person engaged in the business of manufacturing, assembling, or importing of electronic products.

Most radiation-emitting products are not considered to be medical devices. However, if you make any medical claims, your product is a medical device subject to the provisions of the FD&C Act for medical devices in addition to the provisions for radiation-emitting products.

Examples of radiation-emitting electronic products subject to the provisions of the FD&C Act and therefore regulated by the FDA are listed in 21 CFR 1000.15 and include:

1. Ionizing electromagnetic radiation:
   Television receivers
   Accelerators
   X-ray machines (industrial, medical, research, educational)
2. Particulate radiation and ionizing electromagnetic radiation:
   Electron microscopes
   Neutron generators
3. Ultraviolet:
   Biochemical and medical analyzers
   Tanning and therapeutic lamps
   Sanitizing and sterilizing devices
   Black light sources
   Welding equipment
4. Visible:
   White light devices
5. Infrared:
   Alarm systems
   Diathermy units
   Dryers, ovens, and heaters
6. Microwave:
   Alarm systems

Diathermy units
Dryers, ovens, and heaters
Medico-biological heaters
Microwave power generating devices
Radar devices
Remote control devices
Signal generators

7. Radio and low frequency:
   Cauterizers
   Diathermy units
   Power generation and transmission equipment
   Signal generators
   Electromedical equipment
8. Laser:
   Art-form, experimental, and educational devices
   Biomedical analyzers
   Cauterizing, burning, and welding devices
   Cutting and drilling devices
   Communications transmitters
   Range finding devices
9. Maser:
   Communications transmitters
10. Infrasonic:
    Vibrators
11. Sonic:
    Electronic oscillators
    Sound amplification equipment
12. Ultrasonic:
    Cauterizers
    Cell and tissue disintegrators
    Cleaners
    Diagnostic and nondestructive testing equipment
    Ranging and detection equipment

The Code of Federal Regulations (CFR) is a codification of the general and permanent rules published by the Government Printing Office (GPO) in the *Federal Register* (FR) by the executive departments and agencies of the federal government. The CFR is divided into 50 titles that represent broad areas subject to federal regulation. Each title is divided into chapters that usually bear the name of the issuing agency. Each chapter is further subdivided into parts covering specific regulatory areas.

The CDR Title 21 — Food and Drugs list Part 1040 — Performance Standards for Light-Emitting Products is the product performance standard for laser products. The requirements of this section depend on the classification of the laser or laser product.

TABLE 11.1
Classifications of the Laser or Laser Product

Protective housing	Required for all class lasers or laser systems
Safety interlock	Required for all class lasers
Viewing optics	Required for all class lasers
Scanning safeguards	Required for all class lasers
User information	Required for all class lasers
Certification label	Required for all class lasers
Service information	Required for all class lasers
Emission indicator	Required for class 2, 3a, 3b, and 4
Beam attenuator	Required for class 2, 3a, 3b, and 4
Aperture label	Required for class 2, 3a, 3b, and 4
Class warning label	Required for class 2, 3a, 3b, and 4
Product literature	Required for class 2, 3a, 3b, and 4
Protective housing label	Required for class 2, 3a, 3b, and 4
	Depends on level of interior radiation for class 1
Location of controls	Required for class 2, 3b, and 4
Remote control connector	Required for class 3b and 4
Key control	Required for class 3b and 4
Reset	Required class 4 only

Section 1040.10, Laser Products, states that the provisions of this section and Section 1040.11, as amended, are applicable as specified to all laser products manufactured or assembled after August 1, 1976, except when such a laser product is either (a) sold to a manufacturer of an electronic product for use as a component (or replacement) or (b) sold by or for a manufacturer of an electronic product for use as a component (or replacement).

11.5.1 Product Filing

The Office of Compliance (CDRH) receives product submissions. The report must follow the appropriate guide (21 CFR 1002.7). The CDRH does not approve these reports or products. It is the manufacturer's responsibility to certify that their products comply with all applicable standards (21 CFR 1010 - 1050), based on a testing program in accordance with good manufacturing practices. Prior to the shipment of products in interstate commerce, 21 CFR 1002 requires the manufacturer to submit the report and comply with all applicable importation requirements (21 CFR 1005).

Once the report is submitted, the product is legal to sell in the United States. The manufacturer does not have to wait for a response from the CDRH. The submittal package is sent off, and later an accession number (tracking number) is assigned to it and the manufacturer is notified of that number. Sometime after that the manufacturer either receives a letter stating that no additional information is required (meaning the CDRH agrees the product is compliant) or a deficiency letter is sent.

If there are deficiencies, the CDRH may not approve the firm's quality control and testing program or may determine that the product contains a radiation defect or that the product fails to comply with a standard. The CDRH notifies the manufacturer if such a determination is made. The CDRH may require the manufacturer to cease the product's introduction into U.S. commerce until deficiencies are corrected and to initiate a corrective action program (21 CFR 1003 - 1004) for products already introduced into commerce.

11.5.1.1 Special Note

One does not have to place a laser or laser system into commerce to be considered a laser product manufacturer in the eyes of the CDRH. According to Document 9066-MA,

> The Bureau of Radiological Health has been advised previously by the General Counsel that an electronic product, including laser products, constructed on a one-time basis by a particular company for use by that company in its manufacturing process at the place where constructed is not considered "manufacturing." If, however, the products are made on a continuing basis in the course of a commercial enterprise and used by employees other than those directly involved in the manufacture of the electronic product, the company is considered to be engaged in the business of manufacturing products subject to the Act.

In simple words, if those who use the system are not the same staff that made the system, you have become a manufacturer according to the CDRH. An example would be if a marking system is designed within a company and then applied to several assembly lines. According to the CDRH, users of such equipment should have the same assurances that they are dealing with a safe product as anyone who would have purchased a similar commercial system.

11.5.2 Original Electronic Manufacturers

Not every product sold in the United States need be a certified product; it can be an original electronic manufacturer (OEM) product. A laser diode can be an OEM product. These products are considered components, which will go into a laser product. The larger product is certified, and the OEM laser component is addressed as part of that certification application.

11.5.3 Laser Light Shows

The CDRH is also charged with regulating laser light show projectors and laser light shows within the United States if they utilize class 3B or class 4 lasers. A firm desiring to conduct laser light shows must obtain a variance from the CDRH, which is a permit to conduct the show, and agree to follow the CDRH rules and guidelines for such shows.

All manufacturers of class 3B and 4 laser light shows and laser light show projectors must have approved variances from the CDRH to perform laser light shows and introduce laser light show projectors into U.S. commerce. Prior to performing laser light shows or introducing laser light show projectors into U.S. commerce, the manufacturer must submit the following to the CDRH:

1. A product report describing the laser projector
2. A laser light show report describing the laser light show
3. A variance application requesting permission to deviate from the federal laser performance standards

The common venues for laser light shows are outdoor areas, hotel or motel ballrooms or meeting rooms, trade shows or conventions, indoor arenas, planetariums or other dome projection structures, theaters, store displays, discotheques or night clubs, and museums. These shows can involve front screen projections, rear screen projections, holographic displays, multiple reflection and diffraction effects, audience scanning (also includes scanning any accessible uncontrolled areas), reflections from stationary mirrors or mirrored surfaces, stationary irradiation of rotating mirror balls, scanning irradiation of rotating mirror balls, fiber optic projections, and fog, smoke, or other scattering enhancement effects.

The variance can be for a single show or a complete tour schedule. Performance of pre- and postshow checklists is required, although these are generally completed after the show rather than when required. A sample of the control requirements listed in the variance is:

1. All laser products, systems, shows, and projectors will be certified to comply with 21 CFR 1040.10
2. Effects not specifically indicated in this variance application will not be performed. No other effects will be added until an amendment to the variance has been obtained and the required reports or supplements, as applicable, have been submitted
3. Scanning, projection, or reflection of laser and collateral radiation (light show radiation) into audience or other accessible uncontrolled areas will not be permitted except for diffuse reflections produced by the atmosphere, added atmospheric scattering media, and target screens
4. Laser radiation levels in excess of the limits of class 1 will not be permitted at any point less than 3.0 m above any surface upon which persons other than operators, performers, or employees are permitted to stand or 2.5 m below or in lateral separation from any place where such persons are permitted to be. Operators, performers, and employees will not be required or allowed to view radiation above the limits of class 1 or be exposed to radiation above the limits specified in 21 CFR 1040.11(c)
5. Any product that relies on scanning to meet access, exposure, or product class limits will incorporate a scanning safeguard system that directly senses scanner motion and that will react fast enough to preclude exceeding the applicable limit

6. An employee of the variance holder will be responsible for the training and the conduct of the operator
7. Immediately terminate the emission of light show radiation in the event of any unsafe condition, or, for outdoor shows, upon request by any air traffic control officials
8. All laser light shows must be under the direct and personal control of a trained, competent operator
9. The maximum laser projector output power will not exceed the level required to obtain the intended effects
10. The projection system (i.e., the projector and all other components used to produce the lighting effects) will be securely mounted or immobilized to prevent unintended movement or misalignment. Beam masking will be provided as a built-in part of the system design to prevent overfilling of screens, beam stops, targets, and so on
11. A copy of the variance application, the approval letter, current procedures, and records relating to each particular show will be with the operator or other responsible individual and will be made available for inspection by the FDA and other responsible authorities

11.5.3.1 Laser Light Show Notification

The light show operator is required to make a number of notifications in addition to the variance application to the CDRH. The FAA must be notified about any projections into open airspace at any time (including set-up, alignment, rehearsals, performances, etc.). If the FAA objects to any laser effects, the objections are to be resolved and any conditions requested by FAA adhered to. If these conditions cannot be met, the objectionable effects are deleted from the show. State and local radiation control agencies are responsible for all shows to be performed within their jurisdictions. All requirements of state and local law have to be satisfied and any objections raised by local authorities resolved or the effects must be deleted. The notification needs to be in advance as early as possible. It needs to provide a show itinerary, with dates and locations clearly and completely identified, and a basic description of the proposed effects including a statement of the maximum power output intended. A laser light show production may not begin until a variance approval letter that indicates the conditions under which the laser light show may be produced is received.

11.5.3.2 Class 3A and 3R Laser Light Shows

Laser light show regulations only apply to shows using class 3B or class 4 lasers. With current laser technology (frequency-doubled Nd:YAG laser), many venues now use class 3A laser devices. Nightclubs and other venues are excellent candidates for such low-power laser shows. A concern is audience scanning. Without a variance, to review effects, some class 3A scanning may occur.

11.5.3.3 Special Note to LSOs

The majority of laser light shows in the United States do not take place at entertainment venues, but rather at business meetings (annual sales meeting, etc). The LSO needs to confirm that any company putting on a laser light show for your firm has a current variance and is completing required pre- and postshow checklists.

11.5.3.4 CDRH Inspections

The CDRH has a compliance division, and inspections are conducted at laser light shows and there are periodic inspections of laser product manufacturers.

11.5.4 CDRH Databases

11.5.4.1 MAUDE Data (Manufacturer User Facility and Distributor Experience Database)

MAUDE data documents reports of adverse events involving medical devices. The data consists of all voluntary reports since June 1993, user facility reports since 1991, distributor reports since 1993, and manufacturer reports since August 1996.

11.5.4.2 MDR Data (Device Experience Network Database)

MDR data contains information from the CDRH's former database, the device experience network (DEN). The reports include mandatory manufacturer reports on devices that may have malfunctioned or caused a death or serious injury. These reports were received under the mandatory Medical Device Reporting Program (MDR) from 1984 to 1996 and voluntary reports were made up to June 1993. There are over 600,000 reports.

The reports are censored to some degree because the FDA is required under the U.S. Freedom of Information and Privacy Acts (SEC 552, Title 5, USC) (PL 93-579) to delete, prior to public disclosure, any information that constitutes trade secrets and confidential, commercial, or financial information and any personal, medical, and similar information that would constitute a clearly unwarranted invasion of personal privacy. Included in the deletion requirements are all identifications of the reporters of the events.

11.5.5 Medical Products

The primary focus of the CDRH is medical products: ensuring the safety and effectiveness of medical devices and eliminating unnecessary human exposure to man-made radiation from medical, occupational, and consumer products. The center accomplishes its mission by:

1. Reviewing requests to research or market medical devices
2. Collecting, analyzing, and acting on information about injuries and other experiences in the use of medical devices and radiation-emitting electronic products
3. Setting and enforcing good manufacturing practice regulations and performance standards for radiation-emitting electronic products and medical devices
4. Monitoring compliance and surveillance programs for medical devices and radiation-emitting electronic products
5. Providing technical and other nonfinancial assistance to small manufacturers of medical devices

For the CDRH, medical devices are classified into class I, II, and III. Regulatory control increases from class I to class III. The device classification regulation defines the regulatory requirements for a general device type. Most class I devices are exempt from Premarket Notification 510(k), most class II devices require Premarket Notification 510(k), and most class III devices require premarket approval. A description of device classifications and a link to the product classification database can be found at: http://www.fda.gov/cdrh/devadvice/313.html. The basic regulatory requirements that manufacturers of medical devices distributed in the United States must comply with are:

1. Premarket Notification 510(k), unless exempt, or premarket approval (PMA)
2. Establishment registration on form FDA-2891
3. Medical device listing on form FDA-2892
4. Quality system (QS) regulation
5. Labeling requirements
6. Medical device reporting (MDR)

11.5.5.1 Premarket Notification 510(k)-21 CFR Part 807 Subpart E

If a medical device requires the submission of a Premarket Notification 510(k) it cannot be commercially distributed until the manufacturer receives a letter of substantial equivalence from the FDA, authorizing it to do so. A 510(k) must demonstrate that the device is substantially equivalent (a) to one legally in commercial distribution in the United States before May 28, 1976, or (b) to a device that has been determined by FDA to be substantially equivalent. Information on preparing a 510(k) submission can be found at http://www.fda.gov/cdrh/devadvice/314.html.

Most class I devices and some class II devices are exempt from Premarket Notification 510(k) submission. A list of exempt devices is located at http://www.accessdata.fda.gov/scripts/cdrh/cfdocs/cfpcd/315.cfm.

11.5.5.2 Premarket Approval-21 CFR Past 814

Products requiring PMAs are class III devices that are high-risk devices that pose a significant risk of illness or injury or devices found to be not substantially equivalent to class I and II through the 510(k) process. The PMA process includes the submission of clinical data to support claims made for the device. The PMA is an actual approval of the device by FDA. A description of the process and instructions for filing a PMA application can be found at http://www.fda.gov/cdrh/devadvice/pma/.

11.5.5.3 Investigational Device Exemption (IDE)-21 CFR Part 812

Clinical trials using unapproved medical devices on human subjects are performed under an IDE. Clinical studies with devices of significant risk must be approved by the FDA and by an institutional review board (IRB) before the study can begin. Studies using devices of nonsignificant risk must be approved by the IRB before the study can begin.

A description of the IDE process and information on FDA requirements for conducting a clinical study of an unapproved medical device can be found at http://www.fda.gov/cdrh/devadvice/ide/index.shtml.

11.5.5.4 Establishment Registration form FDA-2891-21 CFR Part 807

Manufacturers (domestic and foreign) and initial distributors (importers) of medical devices must register their establishments with the FDA. A description of the establishment registration process and instructions for completion of the establishment registration form FDA-2891 can be found at http://www.fda.gov/cdrh/devadvice/341.html.

Once a year, the FDA sends the registration form FDA-2891(a) to all registered firms to be verified, corrected, and returned by the firm as a yearly registration. In addition to registration, foreign manufacturers must also designate a U.S. agent. Information on U.S. agents can be found at http://www.fda.gov/cdrh/usagent/.

11.5.5.5 Medical Device Listing form FDA-2892-21 CFR Part 807

All medical devices that are manufactured and imported into the United States must be listed with the FDA on medical device listing form FDA-2892. Guidance is available from the CDRH through their Web site.

TABLE 11.2
Center for Devices and Radiological Health Reporting Requirements versus Laser Product Classification

Products	Manufacturer: Product reports Section 1002.10	Manufacturer: Supplemental reports Section 1002.11	Manufacturer: Abbreviated reports Section 1002.12	Manufacturer: Annual reports Section 1002.13	Manufacturer: Test records Section 1002.30 (a)[1]	Manufacturer: Distribution records Section 1002.30 (b)[2]	Dealer and Distributor: Distribution records Sections 1002.40 and 1002.41
Laser products (1040.10, 1040.11)							
Class 1 lasers and products containing lasers[7]	X			X	X		
Class 1 laser products containing class 2a, 2, and 3a, lasers[7]	X			X	X	X	
Class 2a, 2, and 3a lasers and products other than class 1 products containing such lasers[7]	X	X		X	X	X	X
Class 3b and 4 lasers and products containing such lasers[7]	X	X		X	X	X	X

CDRH Website

TABLE 11.3
Sample of CDRH Guidance and Publications

Compliance Guide for Laser Products (FDA 86-8260)
Consolidated Review of Submissions for Lasers and Accessories #G90-1 (blue book memo)
Cosmetic Laser Surgery: A High-Tech Weapon in the Fight Against Aging Skin
Dental Laser Facts
Dental More Gentle with Painless "Drillings" and Matching Fillings
FDA Approves Laser for LASIK
FDA Authorizes Seizure of Unapproved Lasers
FDA Clears First Laser for Treating Tooth Decay
FDA Issues Warning on Misuse of Laser Pointers
Guidance on Electronic Products which Emit Radiation
Guidance on the Content and Organization of a Premarket Notification for a Medical Laser
Guidance on the Department of Defense Exemption from the FDA Performance Standard for Laser Products; Guidance for Industry and FDA

(Continued)

TABLE 11.3
Sample of CDRH Guidance and Publications (*Continued*)

Guide for Preparing Product Reports for Lasers and Products Containing Lasers
Hair Removal Laser Facts
Laser Light Show Safety — Who's Responsibility (FDA 86-8262)
Laser Products-Conformance with IEC 60825-1, Am.2 and IEC 60601-2-22; Final Guidance for Industry and FDA (Laser Notice 50)
Letter to Ophthalmologists about Lasers for Refractive Surgery
Manufacturers and Users of Lasers for Refractive Surgery
Reporting Guide for Laser Light Shows and Displays (21-CFR1002) (FDA 88-8140)
Responsibilities of Laser Light Show Projector Manufacturers, Dealers, and Distributors; Final Guidance for Industry and FDA - Laser Notice 51
Review of Laser Submissions #G88-1 (blue book memo)
Revised Guide for Preparing Annual Reports on Radiation Safety Testing of Laser and Laser Light Show Products (replaces FDA 82-8127)
FDA Approves Laser for LASIK
Lasik Updates
LASIK Eye Surgery
Vision Correction: Taking a Look at What's New

CDRH Website

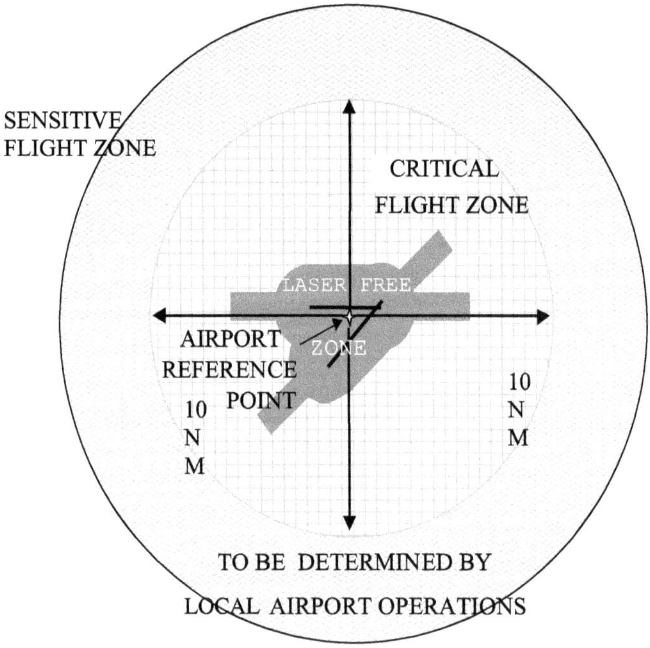

FIGURE 11.1 Sensitive flight zone in nautical miles.

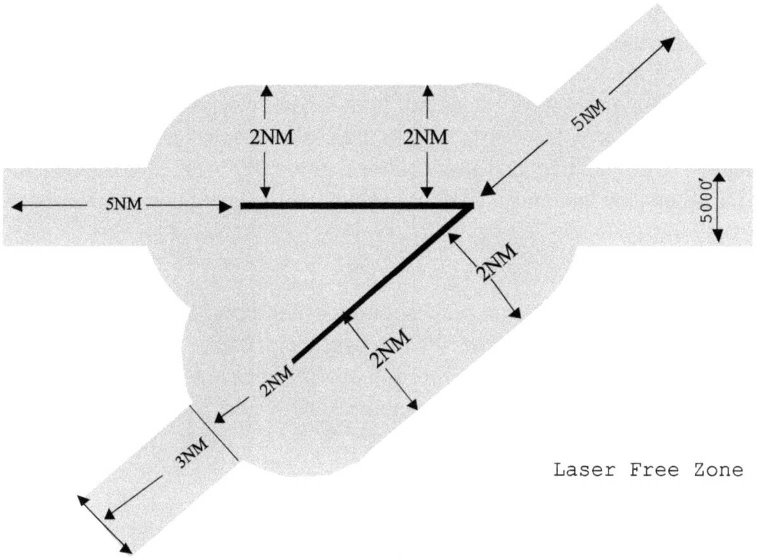

FIGURE 11.2 Runway laser-free zone in nautical miles.

11.6 BRITISH REGULATIONS AND STANDARDS

11.6.1 Standards Versus Legislation

United Kingdom law consists of acts of Parliament and regulations. The latter tend to be more specific and are often "made" under an act. Organizations and individuals can be prosecuted for not complying with the requirements of this legislation. However, it is the courts that generally make the decisions. In the absence of case law, that is, examples of the law being applied in similar situations, interpretation often comes down to common sense and experience of similar situations.

Standards are usually set to ensure that components are manufactured in a particular way or a repeated task is carried out uniformly. Without standards it would be difficult for one manufacturer to make a component fit together with a component from another manufacturer. Standards are agreed upon between interested parties on an international, national, or local basis.

Laser safety standards have evolved over a number of years; the first was published in 1964. Members of the International Electrotechnical Commission (IEC) agree on current standards internationally. These are then adopted by the European Union and issued as Euronorms. The United Kingdom, as a member of the European Union, is then required to adopt these Euronorms as British standards. This is what happened, for example, with BS EN 60825-1:1994.

British standards are not law. The laser standard is relatively unique in that it is a manufacturing standard; that is, a laser product manufactured in compliance with the standards would be expected to have certain features and be classified. The laser standard is also a user safety standard. In the absence of specific acts or regulations covering laser safety, the courts look to available documentation, including British standards. Manufacturers and users can use different means of achieving the same level of safety, but the onus is likely to be on them to prove how this has been done.

11.6.2 Legislation

Two categories of legislation are important when considering laser safety.These are consumer legislation and health and safety legislation. The former is generally enforced by trading standards, through the Department of Trade and Industry, while the latter is enforced by either the Health and Safety Executive or local environmental health officers, depending on the nature of the premises.

11.6.2.1 Consumer Legislation

Consumer legislation is intended to protect purchasers from a wide range of risks, which include risks to health. The parent legislation is the Consumer Protection Act 1987. Like all recent legislation, this can be downloaded from www.legislation.hmso.gov.uk.

Part II of the act covers consumer safety and includes the capability for making regulations under the act. Specific regulations of interest are the General Product Safety Regulations 1994, the Electrical Equipment (Safety) Regulations 1994, and the Medical Devices Regulations 1994. The General Product Safety Regulations apply strictly to products intended for the consumer's private use. However, if a professional product would be expected to be used privately, then it is required to meet the requirements of the regulations.

For this consumer legislation, compliance with the appropriate laser safety standards should be adequate. However, special consideration may need to be given to the assumptions made on use, certainly if the product is intended for domestic use and perhaps by children.

11.6.2.2 Health and Safety Legislation

There is a great deal of health and safety legislation. A useful summary is published by the Health and Safety Executive and is available from HSE Books (*Essentials of Health and Safety at Work*, HSE, 1994, ISBN 0 7176 0716 X). It lists the legislation and gives a brief summary of the requirements for particular subjects. It also explains the role of health and safety inspectors.

11.6.2.3 The Health and Safety at Work Act

The Health and Safety at Work Act 1974 (ISBN 0 10 543774 3) is what is termed an enabling act. This means it is very general but allows for regulations to be made that are more specific. The primary aim of the act is to secure the health, safety and welfare of people at work and other people who may be affected by the work (such as the public).

The act created two bodies: the Health and Safety Commission (HSC) and the Health and Safety Executive (HSE) (Sections 10 to 14). The HSC consists of a full-time chairman and between six and nine part-time members. Its functions include promoting the objectives of the act, carrying out and encouraging research and training, providing an information and advisory service, and putting forward proposals for new regulations.

The major part of the HSE's work is involved with the inspectorates. It also carries out work for the HSC. The HSE's field staff includes the Employment Medical Advisory Service (EMAS), which is responsible for advising on the safeguards and improvement of the health of people at work and people seeking work or training for work.

Sections 2 to 9 of the act place "general duties" on employers, employees, manufacturers, suppliers, and so on. The duties of care are expressed in general terms so that they apply to any work activity imaginable.

11.6.2.3.1 Duties of Employers

The general duties of employers to their employees are found in Section 2 of the act. It is the duty of every employer to ensure, so far as is reasonably practicable (see note below), the health, safety, and welfare at work of all employees (Section 2.1). In addition, the employer must provide safe equipment, safe systems of work, and so on (2.2a) and instruction, training, and supervision (2.2c) as required. Similar considerations apply for any maintenance procedures (2.2d). Section 3 extends these principles to self-employed persons and the public.

11.6.2.3.2 Duties of Manufacturers and Suppliers

Section 6 places similar responsibilities on designers, manufacturers, importers, and suppliers of equipment to ensure that their products are inherently safe through design features and construction. They must also carry out any testing required to prove this, and a key requirement is that the user must be supplied with comprehensive safety and operating information.

11.6.2.3.3 Duties of Employees

Sections 6 and 7 make it an offence for anyone ("employed persons") to willfully interfere with or deliberately tamper with equipment; in other words, all persons have a "duty not to misuse." Employees also have a duty to take reasonable care for their own health and safety and that of others and should cooperate with others, such as safety supervisors, to this end. This was one of the major changes brought in by this act, whereby for the first time employees were legally responsible for their own safety.

11.6.2.4 Regulations and Approved Codes of Practice

Regulations can be made under the act on the basis of proposals from the HSC after a suitable period of consultation. Examples of these are given later.

11.6.2.4.1 Notices

The act provides for the use of notices to require people to take action to improve standards or to prevent accidents. Law backs such notices, and failure to comply may result in prosecution. There are two types of notices: improvement and prohibition. An *improvement* notice is issued when the act or any of its regulations are not being complied with but the risk of serious personal injury is not high. A time limit is placed on the person to comply. A *prohibition* notice means that the work must stop immediately because the inspector believes that serious personal injury may result if the work proceeds. The work need not have already been taking place.

11.6.2.4.2 So Far as is Reasonably Practicable

This term is used widely in current safety legislation as well as in the act. It is not defined, but it has acquired specific meanings, having been tested in the courts. The expression means that the degree of risk in a particular activity or environment can be balanced against the time, trouble, cost, and physical difficulty of taking measures to avoid the risk. If these are so disproportionate to the risk that it would be unreasonable for the persons concerned to have to incur them to prevent it, they are not obliged to do so. The greater the risk, the more likely it is that it is reasonable to go to substantial expense, trouble, and invention to reduce it. It is important to remember that the size or financial position of the employer is not taken into account.

11.6.3 Regulations Under the Act

A number of regulations have been issued under the Health and Safety at Work Act. The following is only a brief summary of the individual requirements.

11.6.3.1 Reporting of Injuries, Diseases, and Dangerous Occurrences Regulations (RIDDOR) 1995

The main purpose of these regulations is to provide information to HSE and to local authorities to assist with accident and disease prevention studies by indicating where and how problems arise and by finding trends.

The types of injuries and dangerous occurrences that need to be reported are included in the regulations. A number of them can result from work involving a laser. If laser radiation is accessible, then the loss of sight in an eye is a possibility. The mechanical aspects, such as moving parts or the weight of the laser, may result in other physical injuries such as fractured bones or amputations.

Electrical shock may be a possibility if maintenance work is being carried out. For these incidents, among others, it is necessary to notify the enforcing authority by the quickest practical means. A report must then be submitted within 10 days. If a person is unable to work for more than 3 days because of an injury resulting from an incident at work, then a written report must be supplied to the enforcing authority within 10 days, even if the injury or dangerous occurrence is not one of those listed.

Forms are available for carrying out the reporting. However, HSE has the right to ask for further information. A guide to RIDDOR is available from HSE Books (ISBN 0 7176 2431 5).

11.6.3.2 Ionizing Radiations Regulations 1999

The nonbeam hazards section highlights the possibility of x-rays generated from the use of lasers, either from thyratron valves or from the laser-workpiece interface. If the voltage on the thyratron exceeds 30 kV or the radiation dose rate exceeds 1 $\mu Sv\ h^{-1}$ at 10 cm, then the equipment is likely to be covered by the regulations.

The basic requirements of the regulations is to notify the HSE of the work with ionizing radiations, appoint a suitably trained radiation protection supervisor, and produce local rules (written procedures) for the work. It is unlikely that the radiation dose rates will be high enough to require attention to many of the other parts of the regulations. However, an assurance that the dose rate is comparable with background radiation levels is worth receiving. The lead protection around thyratrons has been known to be misaligned.

The regulations are accompanied by an Approved Code of Practice and a nonstatutory guidance: L121, from HSE Books, 2000 (ISBN 0 7176 1746 7).

11.6.3.3 Control of Substances Hazardous to Health Regulations 2002 (COSHH)

Many substances can cause harm if they get into the body, and effects may be immediate or delayed. Some lasers contain harmful chemicals, perhaps in the form of laser dyes or gases. Fumes produced by laser material processing are known to present a hazard. The main requirement of these regulations is that you must carry out an assessment of the risk to health from the substances. The supplier of the laser or the chemicals has a duty to assist with this and may even supply an adequate assessment for you to adopt after ensuring that it is applicable to your particular conditions.

A free guide to the COSHH regulations is available from HSE Books: "COSHH a Brief Guide to the Regulations," INDG 136 (1999). This will be updated by HSE in due course.

11.6.3.4 Electricity at Work Regulations 1989

Apart from laser diodes, most lasers operate at high voltages. Therefore, any work done on them will come under the Electricity at Work Regulations. All electrical equipment, wiring installations, generators or battery sets, and everything connected to them must be maintained to prevent danger. This means carrying out checks and inspections and repairing and testing as necessary — how often will depend on the equipment and where it is used.

Access to electrical danger must be prevented by keeping isolator and fuse box covers closed and (if possible) locked, with the key held by a responsible person. Anyone carrying out electrical work must be competent to do so safely.

A number of HSE publications can help in meeting the requirements of these regulations. The *Memorandum of Guidance on the Electricity at Work Regulations 1989* (ISBN 0 7176 1602 9) is particularly useful.

11.6.3.5 The Supply of Machinery (Safety) Regulations 1992

These regulations were made to comply with the European "Machinery Directive." Essentially, machinery for use at work should be assessed against Essential Health and Safety Requirements, which are listed in Schedule 3 of the regulations. The usual hazards are covered, but lasers receive a special mention in Section 1.5.12 of Schedule 3.

Where laser equipment is used, the following provisions should be taken into account:

1. Laser equipment on machinery must be designed and constructed so as to prevent accidental radiation
2. Laser equipment on machinery must be protected so that effective radiation, radiation produced by reflection or diffusion, and secondary radiation do not damage health
3. Optical equipment for the observation or adjustment of laser equipment on machinery must be such that no health risk is created by the laser beams

11.6.3.6 The Health and Safety (Safety Signs and Signals) Regulations 1996

These regulations came into force on April 1, 1996. They require employers to use a safety sign where there is significant risk to health and safety that has not been avoided or controlled by other means, such as engineering controls. The regulations and the associated guidance (L64; ISBN 0-7176-0870-0) stress that safety signs are not a substitute for other control measures.

Safety signs around laser installations may include the yellow laser starburst warning sign, perhaps with an area designation, red prohibition signs restricting access, and blue mandatory signs covering eye protection.

11.6.3.7 "Six Pack"

Much of our legislation now originates in Europe in the form of European directives. A set of six safety-related regulations came into force on January 1, 1993 in compliance with such a directive, and some have been updated since. These are introduced below with some indication of their applicability to laser safety.

11.6.3.7.1 Management of Health and Safety at Work Regulations 1999

These regulations overlap with much other safety legislation. Their basic purpose is to ensure that employers assess the risks to their employees (and to others) and, where appropriate, take steps to reduce or eliminate those risks.

These regulations are covered in more detail in the risk assessment section. A combined approved code of practice and regulations is available: HSE Books, *Management of Health and Safety at Work* (L21; ISBN 0 7176 2488 9).

11.6.3.7.2 Health and Safety (Display Screen Equipment) Regulations 1992

It is possible that a computer may be an integral part of a laser product. If so, whether these regulations apply depends upon the amount of time a person spends at the keyboard. However, the regulations can also be considered to represent good practice.

The regulations require an assessment to be carried out of the area in which the person uses the computer screen, called the workstation. It also specifies certain features, which the computer equipment should have, such as the ability to move and adjust the keyboard and the screen.

A combined document including guidance and regulations is available from HSE: Display Screen Equipment Work, Health and Safety (Display Screen Equipment) Regulations 1992, Guidance on Regulations (L26; ISBN 0 7176 0410 1).

11.6.3.7.3 Provision and Use of Work Equipment Regulation 1998

The regulations were designed to unify and tidy the existing laws covering equipment at work. In theory the regulations make explicit what is already law or what was previously good practice. *Work equipment* is broadly defined to include everything from hand tools to machines of all kind.

Every employer must ensure that any work equipment is suitable, in health and safety terms, for its purpose and that it is indeed used for that purpose. When new equipment is considered, any health and safety risks associated with it or its intended use must also be considered. Equipment must be properly maintained and any associated maintenance log kept up to date.

If any equipment involves a specific risk or risks, then its use, maintenance, and repair must be restricted to suitably trained and authorized persons. All supervisors and users of such equipment must be given adequate health and safety information and, if appropriate, written instructions. These should include safe operating instructions, an assessment of any foreseeable hazards, and contingency plans on how to deal with them. All these issues may be applicable to work with equipment containing lasers.

A number of other sections within the regulations may be applicable to laser work. Regulation 11 deals with protection from dangerous parts of machinery by means of adequate guarding. Regulation 12 covers the prevention or control of exposure to certain specified hazards, such as the unintended discharge of dust, gas or vapor; equipment catching fire; explosion; and so on. This should preferably be through engineering controls rather than PPE or administrative controls.

A combined document including guidance and regulations is available from HSE: *Safe Use of Work Equipment, Provision and Use of Work Equipment Regulations 1998, Approved Code of Practice* (L22; ISBN 0 7176 1626 6).

11.6.3.7.4 Workplace (Health, Safety, and Welfare) Regulations 1992

These regulations apply more to the environment in which the laser is installed than to the laser itself. The workplace must be clean and be maintained, as must equipment such as emergency lighting, fencing, ventilation equipment, and so on. There should be sufficient lighting, which, where possible, should be natural. There should be adequate space around equipment where people are expected to work. Where it is possible to carry out the work sitting down, a seat must be provided (with footrest, if appropriate). The seat should be suitable for the person who is going to sit on it.

The regulations also deal with the domestic aspects of the workplace such as toilet facilities, changing facilities (with appropriate accommodation for clothing), rest facilities, and facilities where employees can eat.

A combined document including guidance and regulations is available from HSE: Workplace Health and Safety, Workplace (Health, Safety, and Welfare) Regulations 1992, Guidance on Regulations (L24; ISBN 0 7176 0413 6).

11.6.3.7.5 Personal Protective Equipment at Work Regulations 1992

The practical aspects of these regulations are dealt with in the section on PPE. However, the main points will be stated here.

PPE should always be considered a last resort. When working with a laser it is much better for access to the laser radiation to be restricted by engineering means than to rely on users to remember to wear protective goggles. If PPE is required, then it must be provided to employees free of charge. The equipment must fit properly and give adequate protection. If more than one piece of PPE is required then different pieces must be compatible.

Before purchase, all PPE must be assessed and found to be suitable for its intended use. This assessment must include a consideration of the risks involved, a statement of the characteristics any equipment must have to be effective, and finally a comparison between the PPE characteristics required and those that are provided. These assessments must be reviewed in the light of any changes.

The PPE must be maintained, cleaned, and replaced as necessary. Storage areas must be provided for when the PPE is not in use. All persons issued PPE must be provided with any necessary information and be trained in their correct use. They must be instructed on the risks the PPE is designed to avoid, the way to use it, and how to maintain it. The employer and employee have the duty to ensure that any necessary PPE is actually used and that it is used correctly. Any loss or breakage of PPE must be reported immediately to the employer.

A combined document including guidance and regulations is available from HSE: Personal Protective Equipment at Work, Personal Protective Equipment at Work Regulations 1992, Guidance on Regulations (L25; ISBN 0 7176 0415 2).

11.6.3.7.6 Manual Handling Operations Regulations 1992

One of the few fatal accidents attributable to laser work was caused by a laser falling on someone. Lasers and the associated equipment can be very heavy. More than a quarter of the accidents reported each year to the enforcing authorities are associated with manual handling. The regulations establish a clear hierarchy of measures to prevent incidents and accidents from manual handling:

1. Avoid hazardous manual handling operations so far as is reasonably practicable. This may be done by redesigning thc task to avoid moving the load or by automating or mechanizing the process.
2. Make a suitable and sufficient assessment of any hazardous manual handling operations that cannot be avoided
3. Reduce the risk of injury from those operations so far as is reasonably practicable. Particular consideration should be given to the provision of mechanical assistance, but where this is not practicable, then other improvements to the task, the load, and the working environment should be explored.

A combined document including the code of practice and regulations is available from HSE: *Manual Handling, Manual Handling Operations Regulations 1992, Guidance on Regulations* (L23; ISBN 0717624153).

11.6.4 British Standards

The main British standards relating to laser safety are BS EN 60825-1:1994 (Part 1) and BS EN 60825-2:1995. There are, however, a number of other standards, which may be relevant to your work with lasers. These are listed here with a summary of their contents.

BS EN 60825-4: 1998: "Safety of laser products. Laser guards." This standard provides some guidance on laser guards as used around laser materials processing machinery.

PD IEC/TR 60825-5: 2003: "Safety of laser products. Manufacturer's checklist for IEC 60825-1." This is aimed at manufacturers to ensure that they do all that is necessary to comply with part 1 of the standard, having already allocated the laser product to a laser class. It is also useful for users, who can use the checklist to audit new equipment for compliance with the standard.

BS IEC TS 60825-6: 1999: "Safety of laser products. Safety of laser products with optical sources exclusively used for visible information transmission to the human eye." Historically (and for safety reasons) it has been rare for people to look down laser beams. Since light emitting diodes (LEDs) are within the scope of the laser safety standards, there are now products included that people are required to stare at. These products include traffic lights and indicator panels.

BS IEC 60825-8: 1999: "Safety of laser products. Part 8: Guidelines for the safe use of medical laser equipment." This is a technical report and, as such, is only a guide. However, many of the principles are useful. It is also the first such document to include the Loughborough University/NRPB risk assessment methodology (in Annex C).

BS EN 60825-12: 2004: "Safety of laser products. Part 12: Safety of free space optical communication systems used for transmission of information."

PD IEC TR 60825-14: 2004: "Safety of laser products. Part 14: A user's guide."

BS EN 60601-2-22: 1996: This standard is a section of the Medical Electrical Equipment Standard, Part 2, Particular Requirements for Safety, Section 2.122: Specification for Diagnostic and Therapeutic Laser Equipment.

BS EN 61040: 1993: "Specification for Power and Energy Measuring Detectors, Instruments and Equipment for Laser Radiation."

BS EN 207: 1999 and 208: 1999: These two standards relate to eyewear. The first relates to normal use (Specification for Filters and Equipment Used for Personal Eye-Protection Against Laser Radiation), and the second to maintenance and adjustment work (Specification for Personal Eye-Protection Used for Adjustment Work on Lasers and Laser Systems).

PD 5304: 2000: This document, "Safety of Machinery," is very detailed and includes much that is relevant to laser safety, such as enclosure interlocks.

BS EN 292: 1991: "Safety of Machinery. Basic Concepts, General Principles for Design."

11.6.4.1 Military Laser Standard

There is a joint services publication (JSP) on laser safety in the Armed Services and Ministry of Defense establishments: JSP 390 "Military Laser Safety." Based on British Standards, this is a practically based document covering many of the problems unique to defense applications.

11.6.5 Guidance Notes

A few guidance documents are produced for particular industries. They are all now some years old and may not cover the full requirements of the current British Standard, BS EN 60825-1. However, they do give good practical advice.

Medical Applications: The Medical Devices Agency has produced "Guidance on the Safe Use of Lasers in Medical and Dental Practice," which was published late in 1995.

Universities: The Committee of Vice Chancellors and Principals (CVCP) has updated (1992) an earlier guidance document. This is part of a general series on safety in universities. The full title is "Safety in Universities: Notes of Guidance. Part 2:1 Lasers."

Printing: The Printing Industry Advisory Committee publishes a number of guidance documents relevant to their industry. The laser one is called "Laser Safety in Printing" and was last produced in 1990.

Entertainment: The HSE has issued guidance on the safe use of lasers in entertainment: HSG95, available from HSE Books.

11.6.6 Laser Safety Officer

How does the LSO fit into all of this legislation, standards, and guidance? First, there is no legal requirement to appoint a person with the title Laser Safety Officer. This is a requirement of the British standard BS EN60825-1: 1994, where class 3B and class 4 lasers and class 3R lasers outside of the 400 to 700 nm wavelength range are used. However, it may be possible to achieve the same end through a different means. Regulation 7 of the Management of Health and Safety at Work Regulations 1999 requires employers to appoint one or more "competent persons" to assist with health and safety. Where the work uses laser radiation, the LSO may fulfill this role.

It is important to consider where the LSO is located in the management structure. If the LSO is in a management or supervisory position, he or she is likely to have executive control over laser safety, that is, to be able to get things done directly. An LSO in the safety department, for example, may only give advice to managers. It is then the responsibility of the managers to carry the advice through.

11.6.7 Obtaining the Documents Cited

Regulations and acts are available for downloading from the Internet at www.legislation.hmso.gov.uk.

12 Laser Safety Calculations*

12.1 INTRODUCTION

In a modern society that seems to be challenged by math, laser safety calculations provide a double-edged dilemma to the laser safety officer (LSO). An understanding of the principles behind calculations is important to the LSO; the actual performance of such calculations today is limited. In the research setting the nominal hazard zone many times is deemed to be the laser room for convenience, and optical density for laser protective eyewear can be provided by vendors. Most LSOs use software to perform safety calculations, but the LSO should be knowledgeable of the calculations involved and capable of recreating the calculations if needed. Laser safety calculation software does offer advantages to the LSO: easily obtained results and a paper document. It is extremely important to be aware that some of the formulas listed in the laser standards are empirically derived from experimental data. This means that the results of biological studies are plotted and the best mathematical fit for a line is determined and, if necessary, offset by a safety factor. The offset safety factor is not uniform across all wavelengths; it can range from a factor of 1 to 10. Therefore, as our understanding of biological effects and processes expands, exposure limits and other aspects of laser safety calculations may change. For the readers to obtain the most accurate information on these factors, they are better off referencing the latest laser American National Standards Institute (ANSI) or International Electrotechnical Commission (IEC) standard.

This chapter has been designed to serve as in introductory text on laser safety calculations. Many more advanced safety calculations can be performed, and readers are encouraged to work through the examples outlined in Appendix B of the ANSI Z136.1 Standard for a more thorough explanation of the formulas affecting laser safety.

In performing a laser safety calculation, the following rules of thumb are very helpful:

1. *Always* calculate the wavelength in micrometers (μm)
2. *Always* calculate distance in centimeters (cm)
3. *Always* calculate time in seconds (sec)
4. *Always* calculate divergence in milliradians (mrad)

Failing to input the numerical values in the correct units in the formulas is one of the more common mistakes made in laser safety calculations.

* Special thanks for this chapter go to Ken Smith of the University of California at Santa Cruz.

12.2 CONVERTING BETWEEN IRRADIANCE AND RADIANT EXPOSURE

It is important to understand the difference between irradiance (E) and radiant exposure (H). Irradiance is measured in watts per square centimeter and is used for continuous wave (CW) lasers. Radiant exposure (H) is used for both single- and multiple-pulse lasers and is measured in joules per square centimeter. You can convert the two using the following formulas:

$$E_{Peak}(Watts) = \frac{H(Joules)}{t(\text{sec})} \qquad H = E \cdot t$$

Key laser specifications not found on laser labels, but found in laser use manuals, are the beam diameter a, the beam divergence ϕ, and the radiant energy Q or radiant power Φ. These values appear in safety calculations, where they have the units of centimeters, radians, joules, and watts, respectively. The central quantities in a laser safety calculation, however, are the radiant exposure H (measured in joules per square centimeter) and the irradiance E (measured in watts per square centimeter). There may also be some confusion depending on the profile of your beam and the choice of beam diameter.

The most common profile of a laser beam is a Gaussian profile. The diameter a of a Gaussian beam can be specified according the $1/e$ or $1/e^2$ point. Laser manufacturers may often use the $1/e^2$ definition since this area encompasses 90% of the total beam energy. However, safety calculations use the $1/e$ diameter, so determine which one you are using. The two diameters have a simple relation: $a\left(\frac{1}{e^2}\right) = \sqrt{2}a\left(\frac{1}{e}\right)$.

Your beam may not have a Gaussian profile. Another common profile is the top hat, in which the beam's radiant exposure is equal for all points within the beam diameter. This mode makes the beam diameter and the irradiance very easy to define. For a beam with a top hat profile, beam diameter a, and radiant power Φ,

$$E_0 = \frac{4\Phi}{\pi a^2}$$

The subscript for E identifies it as the maximum irradiance, that is, at zero distance from the laser's exit port. The radiant exposure is calculated from this by dividing H by the appropriate exposure time or pulse duration.

For a Gaussian beam, the central irradiance is different from the average irradiance over the area within the beam diameter. However, the maximum permissible exposure (MPE) is not determined according to the peak irradiance of the beam; it is determined by averaging the incident power of the beam over an area defined by the *limiting aperture* of the eye. For visible light this limiting aperture is the diameter of a fully dilated pupil, which is 7 mm. For nonvisible radiation other limiting apertures are defined as follows:

Spectral Region	Period of Exposure	Aperture Diameter (mm)
180 to 400 nm	10^{-9} to 0.25	1.0
	0.25 to 3×10^4	3.5
400 to 1400 mm	10^{-9} to 3×10^4	7.0
1400 nm to 100 μm	10^{-9} to 0.25	1.0
	10 to 3×10^4	3.5
100 to 1000 μm	10^{-9} to 3×10^4	11.0

If the diameter of your beam is similar to the limiting aperture for your wavelength, you may calculate the irradiance, assuming your beam has a top hat profile, using the following equation:

$$E_0 = \frac{4\Phi}{\pi\left[\max\left(a, D_f\right)\right]^2}$$

in which Φ is the total radiant power of the beam, a is the beam diameter (measured at $1/e$ of the peak irradiance), and D_f is the limiting aperture at the appropriate wavelength.

12.2.1 The Accessible Emission Limit

The primary measurement of a laser's hazard potential is the accessible emission limit (AEL), which defines the maximum total power of radiation that can be emitted from a laser of a particular class. Assuming a linear additive effect for radiation absorbed by the eye, the minimum irradiance known to cause a biological effect is converted into a power level for the length of time defined by a given class.

As AELs are mainly used for classifying lasers, they are not immediately useful to a user who wants to know if a particular setup is safe. However, you can use the classification scheme to help make very simple decisions. For instance, if you always keep your power level below that required of a class 2 device, you can be assured that accidental exposure to the beam will not be hazardous. For visible lasers, this means keeping the average power below 1 mW for start-up or alignment procedures.

12.2.2 Maximum Permissible Exposure (MPE)

The MPE should be thought of as the level of brightness that is safe* to view (or be exposed to) for that period of time.

MPE calculations are perhaps the most fundamental and useful calculations to be performed by the LSO and will form a comparative basis to determine what is and is not safe. MPE can be thought of as a threshold limit value (TLV) for a given

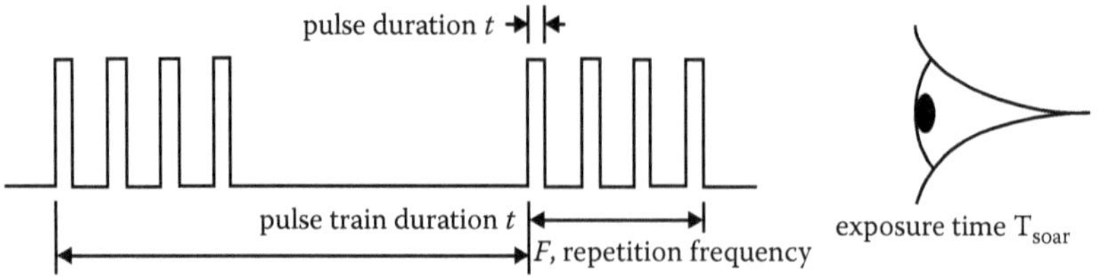

FIGURE 12.1 Explanation of pulse laser terms.

exposure time. Since many factors affect the MPE, a set of values cannot be listed in a table as they are for chemicals or other physical hazards. The target organ (eye or skin), the source size (small or extended), the wavelength, the exposure duration, the type of laser (CW, single, or multiple pulsed), the pulse repetition frequency (PRF), and the pulse length are all factors that can affect the MPE.

12.2.3 Repetitively Pulsed Lasers

The most common lasers in research today are pulsed lasers, because of their higher peak energies. Because of its high peak intensity, the MPE from a pulsed laser is more complicated than for an equivalent continuous source. To determine the correct MPE for a given train of pulses you must know the pulse repetition frequency (F), the duration of a single pulse (t), the total duration of a train of pulses (T), and the total exposure time T_{max} (Figure 12.1).

Three rules limit the MPE per pulse for a train of laser pulses:

1. The MPE per pulse is limited to the MPE for any single pulse (single pulse limit).
2. The MPE per pulse is limited to the MPE for all exposure times between T and T_{max}, divided by the number of pulses n during that time period (average power limit).
3. The MPE per pulse is limited to the MPE for a single pulse multiplied by $n^{-1/4}$, where n is the number of pulses that occur during the period of exposure T_{max} (repetitive pulse limit).

Above a critical frequency (>55 kHz for visible lasers), rule 2 gives the lowest value. In other cases all three limits should be checked.

12.2.4 Extended Source Viewing

If a laser beam has a high divergence, or is viewed at a close distance relative to the beam diameter, the image the beam forms upon the retina may not be a point. In this case, the laser is an extended source and the MPE should be corrected accordingly. Extended source MPEs are applied only in the spectral region of 400 to 1400 nm and where the angle subtended by the eye to capture the entire beam is greater than a minimum angle specified for differing exposure times (Table 12.1).

TABLE 12.1
Extended Source Factors Specified for Exposure Times

Exposure Time (sec)	Angle α_{min} (mrad)
0.7	1.5
0.7 – 10	$2t^{3/4}$
10	11

If the laser beam subtends an angle greater than the minimum for particular viewing circumstances, then the MPE should be multiplied by the factor α/α_{min}. Extended source viewing may apply in cases such as viewing a beam after it passes through an optic or viewing a diffuse reflection at a close distance. However, since an extended source correction factor increases the MPE, it is generally sufficient to perform your calculations simply assuming a point source (worst case scenario).

12.2.5 Diffuse Reflections

The major feature of class 4 lasers such as those found in research laboratories is that even diffuse reflections can be hazardous. Thus, it is important to understand how to calculate the power of a diffusely reflected laser beam. The flux of energy through a point at a given distance and angle from a diffusely reflecting surface is

$$H = \frac{\rho_\lambda Q \cos\theta_v}{\pi r^2}$$

where ρ_λ is the reflectance at the wavelength in question, Q is the energy of the incident beam, θ_v is the angle of observation with respect to the normal, and r is the distance from the point of reflection to the point of observation. For safety calculations, this equation can be greatly simplified by assuming $\rho_\lambda = \cos\,\theta_v = 1$.

12.2.6 The Nominal Hazard Zone

The other major definition for laser safety calculations is the nominal hazard zone (NHZ). This is a distance within which the irradiance of a beam is greater than the MPE; in other words, the area around your laser within which the beam exceeds the MPE. Besides being specific to a given wavelength and time of exposure, a different NHZ can be defined for the beam's path to your eye — direct viewing, specular reflectance, or diffuse reflectance.

Once you have determined the MPE for a specific laser, you can determine the NHZ. To do this you need to know how your beam diameter changes with distance. A Gaussian beam changes its diameter according to a hyperbolic function, with the minimum diameter occurring at the *beam waist*. However, for safety calculations you can use one of two approximations. For lasers in which the beam waist is at or in front of the exit port, the diameter D of a Gaussian beam at a given distance r is given by the following formula:

$$D = \sqrt{a^2 + (r - r_0)^2 \phi^2}$$

where a is the beam diameter at the exit port and ϕ is the beam divergence. Depending on the situation, you may further simplify this equation by dropping the r_0 (waist close to exit port), a (large distance r), or the whole thing (for a large diameter, low divergence, and short distance; D a). In cases where the beam waist is behind the exit port, it may be more appropriate to use a linear approximation:

$$D = a + r\phi$$

12.3 OPTICAL DENSITY

Optical density (OD) is a filter transmission factor, similar to SPF in sun block. The higher the OD number, the higher the attenuation of the material, whether it is used for laser protective glasses or protective windows. It is a simple log ratio of the incident beam versus the transmitted beam. When it comes to laser safety eyewear, the OD most often calculated is for intrabeam or direct exposure. This yields an OD for full protection — complete absorption or blockage of the laser beam.

When one talks about alignment eyewear, we use an OD that allows visualization of the beam and not complete blockage. This is discussed further in Chapter 8. Based upon these typical exposure conditions, the OD required for suitable filtration can be determined. OD is a logarithmic function defined by

$$\text{OD} = \log 1/\text{T}_\lambda$$

$$\text{T}_\lambda = \text{output/input}$$

Therefore, OD = log [input/output] H_0 = input.

$$OD = \log_{10} \frac{H_0}{MPE}$$

where H_0 is the anticipated exposure (joules per square centimeter or watts per square centimeter). MPE is expressed in the same units as H_0.

For full protection results, the OD must be such that it reduces the transmission of the effective wavelength to less than the MPE for the laser in question (Table 12.2). For example, the minimum optical density at the 0.514-μm argon

TABLE 12.2
Logarithmic Values

Log	Factor
1	10
2	100
3	1,000
4	10,000
5	100,000
6	1,000,000
7	10,000,000

laser wavelength for a 600-sec direct intrabeam exposure to the 5-W maximum laser output can be determined as follows:

$$f = 5 \text{ Watts}$$

$$MPE = 16.7 \text{ W/cm}^2 \text{ (using 600-sec criterion)}$$

$$d = 16.7 \text{ W/cm}^2 \text{ (using 600-sec criterion)}$$

$$H_o = [\text{power/area}] = f/A = 4f/\ pd^2$$

$$= [(4)(5.0)/\ p(0.7)^2]$$

$$= 12.99 \text{ W/cm}^2$$

$$OD = \log_{10} [(12.99)/(16.7 \times 10^{-6})]$$

$$OD = 5.9$$

12.4 CHOOSING EXPOSURE DURATION TIMES

There is logic behind the exposure duration time chosen for laser calculations.

12.4.1 Visible Wavelengths (400 to 700 nm): 0.25 sec

The human eye has an aversion to bright light. This aversion is sometimes referred to as the blink response (closing of the eye lid) or aversion reflex (turning of the head). Thus, it is difficult for the human eye to stare at bright light. This becomes the first line of defense for unexpected exposure to some lasers and is the basis of the class 2 concept.

12.4.2 Near Infrared Wavelengths (700 to 1400 nm): 10 sec

The time period chosen by the ANSI Z 136.1 committees represents the optimum worst-case time period for ocular exposures to infrared (principally near infrared) laser sources. It was argued that natural eye motions dominate for periods longer than 10 sec.

12.4.3 Diffuse Reflections (Invisible Wavelengths): 600 sec

This time period is suggested for visible and invisible diffuse reflections during tasks such as alignment.

12.4.4 Ultraviolet Wavelengths: 30,000 sec

This time period represents a 1-day (8-h) occupational exposure. It is the number of seconds in 8 hours (8 h × 60 min/h × 60 sec/min = 28,800 sec). Rounded off, it becomes 30,000 seconds. A limited number of laser users might be exposed to ultraviolet (UV) wavelengths for 8 h a day. Therefore, in the majority of UV wavelengths exposure calculations the LSO will choose a time for that the individual might be exposed to UV wavelengths.

12.4.5 Limiting Aperture Note

All the power that enters the limiting aperture approach will simplify calculations. However, for exposures to large-diameter laser beams or exposures at a great distance from the aperture where the divergence of the beam has expanded its size, this calculation is too conservative to be useful.

Note: This is the approach that some commercially available software uses. It is acceptable when the diameter of the laser is approximately the same size as the limiting aperture or smaller.

Before we can understand this point, we need to review an important concept in measuring the laser beam's diameter. Almost all lasers do not emit a constant intensity over the entire diameter of the laser beam. Most lasers operate in TEM_{00} (transverse electromagnetic) mode, otherwise known as a Gaussian beam profile. This beam profile differs from a constant intensity (or so-called top-hat) profile by the following formula:

$$I = I_0 e^{-\left(\frac{2R^2}{a^2}\right)}$$

It is important to note that the edges of laser radiation of a Gaussian beam extend beyond the diameter indicated. Because there is no sharp edge to use in measuring the size of the beam, manufacturers commonly determine the point where the power emitted equals e^{-2} (or $1/e^2$) as the beams circumference. This is cited because 86.5% of the beam's power lies within this circle. The ANSI standard, however, uses the 1/e point (encompassing 63% of the beam's power).

Therefore, the equation below must be used to convert the manufacturer's specifications before making any calculations.

$$a_{1/e} = \frac{a_{1/e^2}}{\sqrt{2}}$$

13 Nonbeam Hazards

Do not develop tunnel vision; there is more in a laser lab than just lasers.

13.1 INTRODUCTION

While beam hazards (exposure to the laser beam) are the most prominent laser hazards, other hazards pose an equal or possibly greater risk of injury or death. To date, we have not melted anyone with a laser beam, but we have had laser users electrocuted and patients die from operating room fires with the laser as the heat or ignition source. As laser applications expand further into our socicty, a greater number of associated, or nonbeam, hazards will need to be considered to maintain a safe workplace. Here are two examples: at a construction site, which is the greater hazard to the worker — the alignment laser used in a trench or a lack of adequate bracing of the trench, which could cave in and cause a worker's death? When laying an optical fiber for communications, should one be more concerned about confined work in a manhole or the laser fiber? Clearly, one must always be aware of hazards associated with an activity. The number-one associated hazard with laser use is the possible electrical hazard.

The long list of associated hazards can be broken down into hazard categories: physical, chemical, biological, mechanical, and ergonomic human factors. The laser safety officer (LSO) need not be an expert in these areas, but needs to be aware that such hazards exist and be alert for them. Once determined, the appropriate safety specialists should perform a proper evaluation. In some cases users have grown so accustomed to the hazards around them that such hazards may not even occur to them if they are asked to list workplace hazards. An example might be small quantities of optical cleaning solvents.

13.1.1 A Listing of Nonbeam Hazards

A. Physical
1. Noise
2. Vibration
3. Incoherent radiation
4. X-rays
5. High temperature
6. Low temperature
7. Electricity

B. Chemical
1. Toxic substances
2. Carcinogenic substances

3. Irritant substances
4. Dust and particulates
5. Fire

C. Biological

1. Microbiological organisms
2. Viruses

D. Mechanical

1. Trailing cables and pipes
2. Sharp edges
3. Moving parts
4. High-pressure water

E. Ergonomic

1. Workstation layout
2. Manual handling
3. Person-machine interface
4. Ease of operation
5. Shift patterns

13.2 ELECTRICAL HAZARDS

Most lasers contain high-voltage power supplies and often-large capacitors or capacitor banks that store lethal amounts of electrical energy. In general, systems that permit access to components at such lethal levels must be interlocked. However, during maintenance, service, and alignment procedures, such components often become exposed or accessible. This has caused numerous serious and fatal shocks.

13.2.1 Good Practice Guidelines for Electrical Hazards

1. All equipment should be installed in accordance with recognized national electrical codes.
2. All electrical equipment should be treated as if it were "live."
3. Working with or near live circuits should be avoided. Whenever possible, unplug the equipment before working on it.
4. A buddy system should be used when work on live electrical equipment is necessary, particularly after normal work hours or in isolated areas. Ideally, the person should be knowledgeable of first aid and CPR.
5. Rings and metallic watchbands should not be worn, nor should metallic pens, pencils, or rulers be used while one is working with electrical equipment.
6. Live circuits should be worked on using one hand when it is possible to do so.
7. When one is working with electrical equipment, only tools with insulated handles should be used.

8. Electrical equipment that gives the slightest perception of current when touched should be removed from service, tagged, and repaired prior to further use.
9. When working with high voltages, consider the floor conductive and grounded unless standing on suitably insulated dry matting normally used for electrical work.
10. Live electrical equipment should not be worked on when one is standing on a wet floor or when the hands, feet, or body are wet or perspiring.
11. Do not undertake hazardous activities when fatigued, emotionally stressed, or under the influence of medication that dulls or slows the mental and reflex processes.
12. No one should work on lasers or power supplies unless qualified and approved to perform the specific tasks.
13. Before working with electrical equipment, de-energize the power source. Lock and tag out the disconnect switch.
14. Check that each capacitor is discharged, shorted, and grounded before working near capacitors.
15. When possible, use shock-preventing shields, power supply enclosures, and shielded leads in all experimental or temporary high-voltage circuits.
16. Follow lockout/tagout procedures when working with hard-wired equipment.

13.2.2 Potential Electric Hazard Problems

1. Uncovered electrical terminals
2. Improperly insulated electrical terminals
3. Hidden power off/on warning lights
4. Lack of personnel training
5. Buddy system not being practiced during maintenance and alignment work
6. Non-earth-grounded or improperly grounded laser equipment
7. Excessive wires and cables on the floor that create fall and trip hazards

What follows is additional information on some of the more common non-beam hazards you might encounter.

13.3 PHYSICAL HAZARDS

13.3.1 Cryogenic Fluids

Cryogenic liquids (especially liquid nitrogen) may be used to cool the laser crystal and associated receiving and transmitting equipment. These liquefied gases are capable of producing skin burns, and as they evaporate, they replace the oxygen in the area; this is a particular danger in small unventilated rooms. Adequate ventilation must be ensured. The storage and handling of cryogenic liquids must be performed in a safe manner. Insulated handling gloves that can be quickly removed should be worn. Clothing should have no pockets or cuffs to catch spilled cryogenics. Suitable eye protection must be worn. If a spill occurs on the skin, flood the skin contact area with large quantities of water. Cryogenic fluids are potentially explosive when ice collects in valves or connectors that are not specifically designed for use with cryogenic fluids. Condensation of oxygen in liquid nitrogen presents a serious explosion hazard if the liquid oxygen comes in contact with any organic material. All cryogenic liquids should be used with caution because of the potential for skin or eye damage from the low temperature and the hazards associated with pressure buildups in enclosed piping or containers. Portable containers should only be used where there is sufficient ventilation. Do not place containers in a closet or other enclosed space where there is no ventilation supply to the area. The buildup of inert gas in such an area could generate an oxygen-deficient atmosphere.

A full face shield, loose fitting cryogenic handling gloves, apron, and cuffless slacks are recommended for transferring cryogenic fluids. Special vacuum jacket containers with loose-fitting lids should be used to handle small quantities. Vacuum jacketed containers should have overpressure relief devices in place. When plumbing cryogenic liquids, it is very important to include a pressure relief valve between any two shutoff valves. Also, any space where cryogenic fluids may accumulate (consider leakage into enclosed equipment as well) must be protected by overpressure relief devices. Tremendous pressures can be obtained in enclosed spaces as the liquid converts to gas. For example, 1 cm^3 of liquid nitrogen will expand to 700 times this volume as it converts (warms) to its gaseous state. Lines carrying liquid should be well insulated. Containers to be filled with cryogenic liquids should be filled *slowly* to avoid splashing. Cryogenic containers showing evidence of loss of vacuum in their outer jacket (ice buildup on the outside of the container) should not be accepted from the gas supplier. Contact with air (or gases with a higher boiling point) can cause an ice plug in a cryogenic container. Should ice plugs be noted, obtain assistance.

13.3.2 Radio Frequency (RF)

Some lasers contain RF-excited components such as plasma tubes and Q switches. Unshielded and loose-fitting components may generate RF fields. RF leakage surveys should be conducted each time RF cables are reconnected. The appropriate protection guide for RF and microwave energy is that given in the American National Standard "Safety Levels with Respect to Human Exposure to Radio Frequency Electromagnetic Fields, 300 kHz to 100 GHz" (American National Standards Institute [ANSI] C95.1).

13.3.3 Plasma Emissions

When high-power pulsed laser beams (peak irradiance of the order of 10^{12} W/cm^2) are focused on a target, plasma is generated that may also emit collateral radiation. The plasma may contain hazardous "blue light" ultraviolet (UV) emissions.

13.3.4 UV and Visible Radiation

When targets are heated to very high temperatures, as in laser welding and cutting, an intense bright light is emitted. This light often contains large amounts of short-wavelength or blue light, which may cause conjunctivitis, photochemical damage to the retina, and erythema (sunburn-like reactions) on the skin. Other sources of UV and visible radiation are laser discharge tubes and pump lamps. The levels produced may be eye and skin hazards. They need to be suitably shielded to reduce exposure to levels below the ANSI Z 136.1 (extended source), the Occupational and Safety Health Administration (OSHA) PELs, and/or ACGIH threshold limit values.

13.3.5 Explosion Hazard

Lasers and ancillary equipment may present explosion hazards. High-pressure arc lamps and filament lamps used to excite the lasing medium must be enclosed in housings that can withstand an explosion if the lamp disintegrates. In addition, the laser target and elements of the optical train may shatter during laser operation and should be enclosed in a suitable protective housing. Capacitors may explode if subjected to voltages higher than their rating and must be adequately shielded; it is recommended that capacitors be equipped with current-limiting devices. High-energy capacitors should be enclosed in one-eighth-inch thick steel cabinets.

13.3.6 Ionizing Radiation (X-Rays)

X-rays can be produced from two main sources: (a) high-voltage vacuum tubes of laser power supplies, such as rectifiers and thyratrons, and (b) electric discharge lasers. Any power supplies that require more than 15 kV may produce enough x-rays to be a health concern. Although most laser systems use voltages less than 8 kV, some research models may operate above 20 kV.

13.3.7 Noise

Noise levels in laser areas rarely are a concern, outside of headaches from the constant pinging sounds from some pulse lasers onto targets. Noise can exceed safe limits because of high-voltage capacitor discharges and in laser peening operations. In these cases hearing protection may be required. A noise-filled environment can mask the sound of alarms; in such cases visual alarm systems are advisable. An industrial hygienist should perform a noise evaluation if concerns exist.

13.4 CHEMICAL HAZARDS

13.4.1 Laser Dyes

Laser dyes and associated solvents are often composed of toxic and carcinogenic chemicals dissolved in flammable solvents. This creates the potential for fires, chemical spills, and personnel exposures above permissible limits. Frequently, the most hazardous aspect of a laser operation is the mixing of chemicals that make up the laser dye. Dye lasers normally use a lasing medium composed of a complex fluorescent organic dye dissolved in an organic solvent. The use of dimethylsulfoxide (DMSO) as a solvent for cyanide dyes in dye lasers is of particular concern and should be discouraged. DMSO aids in the transport of dyes into the skin.

13.4.1.1 Recommended Guidelines for Laser Dye Use

1. All dyes must be treated as hazardous chemicals. Most solvents suitable for dye solutions are flammable and toxic by inhalation or skin absorption.
2. Obtain material safety data sheets (MSDSs) for all dyes and solvents.
3. Use and store all dyes and solvents in accordance with your institution's chemical hygiene plan if one exists.
4. Prepare and handle dye solutions inside a chemical fume hood.
5. Wear a lab coat, eye protection, and gloves. Contact an industrial hygienist for assistance in glove selection.
6. Pressure-test all dye laser components before using dye solutions. Pay particular attention to tubing connections.
7. Install spill pans (secondary containment) under pumps and reservoirs. While plastic trays maybe economical and available, they are flammable. A superior choice for secondary containment is metal trays.
8. Be alert to contaminated parts.
9. Keep dye mixing areas clean.
10. Tie down tubes so that if they disconnect, one does not have an angry snake on one's hands.

13.4.2 Compressed and Toxic Gases

Hazardous gases (e.g., fluorine, hydrogen chloride) may be used in such laser applications as excimer lasers. If hazardous gases are used, standard operating procedures should contain references for the safe handling of compressed gases, such as seismic restraints, use of gas cabinets, proper tubing and fittings, and so on. The use of mixtures with inert gases, rather than the pure gases, is generally preferred. Hazardous gases should be stored in appropriately exhausted enclosures, with the gases permanently piped to the laser using the recommended metal tubing and fittings. An inert gas purge system and distinctive coloring of the pipes and fittings is also prudent. Compressed gas cylinders should be secured from tipping. Standard compressed gas safety rules should be followed. If questions arise, consult your compressed gas safety specialists.

13.4.3 Fumes, Vapors, and Laser-Generated Air Contaminants from Beam-Target Interaction

Air contaminants may be generated when certain class 3B and class 4 laser beams interact with matter. When the target irradiance reaches a given threshold (approximately 10^7 W/cm^2), target materials, including plastics, composites, metals, and tissues, may vaporize, creating hazardous fumes or vapors that may need to be captured or exhausted. The *plume* is the cloud of contaminants created when there is an interaction between the beam and the target matter. The plume may contain metallic aerosols dust, chemical fumes, and aerosols containing biological contaminants. Some specific examples follow:

1. Ozone produced around flash lamps and concentrations of ozone that build up with high-repetition rate lasers
2. Asbestos fibers released from the firebricks used as backstops for carbon dioxide lasers
3. Polycyclic aromatic hydrocarbons from mode burns on poly(methyl methacrylate) type polymers
4. Hydrogen cyanide and benzene from cutting aromatic polyamide fibers
5. Fused silica from cutting quartz
6. Heavy metals from etching
7. Benzene from cutting polyvinyl chloride
8. Cyanide, formaldehyde, and synthetic and natural fibers associated with other processes

Special optical materials used for far infrared windows and lenses have been the source of potentially hazardous levels of airborne contaminants. For example, calcium telluride and zinc telluride burn in the presence of oxygen when beam irradiance limits are exceeded. Exposure to cadmium oxide, tellurium, and tellurium hexafluoride should also be controlled.

Exposure to these contaminants must be controlled to reduce exposure below acceptable OSHA or national permissible exposure limits. The MSDS may be consulted to determine exposure information and permissible exposure limits. In general, three major control measures are available: exhaust ventilation, respiratory protection, and isolation of the process. Whenever possible, reticulation of the plume should be avoided. Exhaust ventilation, including use of fume hoods, should be used to control airborne contaminants.

13.5 FIRE HAZARDS

Class 4 laser systems represent a fire hazard along with some focused class 3B lasers. Enclosure of class 3 laser beams can result in potential fire hazards if enclosure materials are likely to be exposed to irradiances exceeding 10 W/cm^2. The use of flame retardant materials is encouraged. Opaque laser barriers (e.g., curtains) can be used to block the laser beam from exiting the work area during certain operations. While these barriers can be designed to offer a range of protection, they normally cannot withstand high irradiance levels for more than a few seconds without some damage, including the production of smoke, open fire, or penetration. Users of commercially available laser barriers should obtain appropriate fire prevention information from the manufacturer. Operators of class 4 lasers should be aware of the ability of unprotected wire insulation and plastic tubing to ignite from intense reflected or scattered beams, particularly from lasers operating at invisible wavelengths. Class 4 lasers can also ignite gas atmospheres and cases exist of class 3B lasers igniting a gas environment, when particulates in the gas environment absorb the laser radiation and get hot enough to ignite the gas atmosphere.

13.6 ERGONOMICS AND HUMAN FACTORS

Ergonomic problems can arise from a laser operation that causes awkward arm and wrist positions. If these positions occur for prolonged periods of time, medical problems such as repetitive strain injuries may arise. Back injuries can occur from stretching or strains caused by poorly designed enclosures that are repeatedly taken on and off.

13.7 SEISMIC SAFETY

In some parts of the world, natural disaster conditions such as earthquake need to be considered in the design and use of equipment. Examples would be fastening electronic racks to the floor or walls, and having at least two locking wheels on rolling racks and tying down computer monitors. When possible, bolt down heavy laser equipment. One should be aware of tall objects (bookcases, optical storage racks) that if tipped over would block access into or out of a work space.

13.8 AREA ILLUMINATION

Adequate lighting is a standard recommendation. If low light levels are required, have luminescent strips or arrows to show the way to exits and emergency equipment.

13.9 MECHANICAL HAZARDS

13.9.1 Robotics

With the increasing use of robotics in a variety of settings, robotic safety cannot be overlooked. A common error is to assume the robot is in the off position when it is in a pause mode. Robotic hazards can occur from a number of sources: overloading on the robot, effects from RF sources, software errors, corrosive atmosphere effects on material, and metal stress. The majority of robot laser tools or applications present little hazard from the laser beam. The wide variety of robotic uses presents a broad spectrum of hazard situations. Here is some food for thought when selecting or thinking about robot safeguarding systems. An effective robotic safeguarding system should be based upon a hazard analysis of the robot system's use, programming, and maintenance operations. Among the factors to be considered are the tasks a robot will be programmed to perform, programming procedures, environmental conditions, location and installation requirements, possible human errors, scheduled and unscheduled maintenance, possible robot and system malfunctions, normal mode of operation, and all personnel functions and duties.

An effective safeguarding system protects not only operators but also any others who might work on the robot or integrated equipment and those in the surrounding area. A combination of safeguarding methods may be used. Redundancy and backup systems are especially recommended, particularly if a robot or robot system operates in hazardous conditions or handles hazardous materials. The safeguarding devices employed should not themselves constitute or act as a hazard or curtail necessary vision or viewing by human operators.

13.9.1.1 Robot Accidents

Robotic incidents can be grouped into four categories:

Impact or Collision Accidents: Unpredicted movements, component malfunctions, or unpredicted program changes related to the robot's arm or peripheral equipment can result in contact accidents.

Crushing and Trapping Accidents: A worker's limb or other body part can be trapped between a robot's arm and other peripheral equipment, or the individual may be physically driven into and crushed by other peripheral equipment.

Mechanical Part Accidents: The breakdown of the robot's drive components, tooling or end-effector, peripheral equipment, or its power source is a mechanical accident. The release of parts, failure of the gripper mechanism, or failure of end-effector power tools (e.g., grinding wheels, buffing wheels, deburring tools, power screwdrivers, and nut runners) are a few types of mechanical failures.

Other Accidents: Other accidents can result from working with robots. Equipment that supplies robot power and control represents potential electrical and pressurized fluid hazards. Ruptured hydraulic lines could create dangerous high-pressure cutting streams or whipping hose hazards. Environmental accidents from arc flash, metal spatter, dust or electromagnetic or RF interference can also occur. In addition, equipment and power cables on the floor present tripping hazards.

13.10 SUPPLEMENTAL ELECTRICAL SAFETY SECTION

13.10.1 Life-Threatening Effects

Charles F. Dalziel, Ralph Lee, and others have established the following criteria for the lethal effects of electric shock:

1. Currents in excess of a human's "let-go" current ($\geq$16 mA at 60 Hz) passing through the chest can produce collapse, unconsciousness, asphyxia, and even death.

2. Currents ($\geq$30 mA at 60 Hz) flowing through the nerve centers that control breathing can produce respiratory inhibition, which could last long after interruption of the current.
3. Cardiac arrest can be caused by a current greater than or equal to 1 A at 60 Hz flowing in the region of the heart.
4. Relatively high currents (0.25 to 1 A) can produce fatal damage to the central nervous system. Currents greater than 5 A can produce deep body and organ burns, substantially raise body temperature, and cause immediate death.
5. Electricity flowing through the human body can shock, cause involuntary muscle reaction, paralyze muscles, burn tissues and organs, and kill. The typical effects of various electric currents flowing through the body on the average 150-lb male and 115-lb female body are given in Table 13.1.

13.10.2 Burns

Although a current may not pass through vital organs or nerve centers, internal electrical burns can still occur. These burns, which are a result of heat generated by current flowing in tissues, can be either at the skin surface or in deeper layers (muscles, bones, etc.) or both. Typically, tissues damaged from this type of electrical burn heal slowly. Burns caused by electric arcs are similar to burns from high-temperature sources. The temperature of an electric arc, which is in the range of 4,000 to 35,000°F, can melt all known materials, vaporize metal in close proximity and burn flesh and ignite clothing at distances up to 10 ft from the arc.

13.10.3 Delayed Effects

Damage to internal tissues may not be apparent immediately after contact with the current. Internal tissue swelling and edema are also possible.

13.10.4 Critical Path

The critical path of electricity through the body is through the chest cavity. Current flowing from one hand to the other, from a hand to the opposite foot, or from the head to either foot will pass through the chest cavity, paralyzing the respiratory or heart muscles, initiating ventricular fibrillation, and/or burning vital organs.

13.10.5 Influential Variables

The effects of electric current on the human body can vary depending on the following:

1. Source characteristics (current, frequency, and voltage of all electric energy sources)

TABLE 13.1
Effects of Electric Current on the Human Body

Effect/Feeling	Direct Current (mA) 150 lb	Direct Current (mA) 115 lb	Alternating Current (mA) 60 Hz 150 lb	Alternating Current (mA) 60 Hz 115 lb	Alternating Current (mA) 10,000 Hz 150 lb	Alternating Current (mA) 10,000 Hz 115 lb	Incident Severity
Slight sensation	1	0.6	0.4	0.3	7	5	None
Perception threshold	5.2	3.5	1.1	0.7	12	8	None
Shock, not painful	9	6	1.8	1.2	17	11	None
Shock, painful	62	41	9	6	55	37	Spasm, indirect injury
Muscle clamps source	76	51	16	10.5	75	50	Possibly fatal
Respiratory arrest	170	109	30	19	180	95	Frequently fatals
≥0.03-sec ventritular fibrilation	1300	870	1000	670	1100	740	Probably fatal
≥3-sec ventricular fibrilation	500	370	100	67	500	340	Probably fatal
≥5-sec ventricular fibrilation	375	250	75	50	375	250	Probably fatal
Cardiac arrest	-	-	4000	4000	-	-	Possibly fatal
Organs burn	-	-	5000	5000	-	-	Fatal if it is a vital organ

2. Body impedance and the current's pathway through the body
3. How environmental conditions affect the body's contact resistance
4. Duration of the contact

13.10.5.1 Source Characteristics

An alternating current (ac) with a voltage potential greater than 550 V can puncture the skin and result in immediate contact with the inner body resistance. A 110-V shock may or may not result in a dangerous current, depending on the circuit path, which may include the skin resistance. A shock greater than 600 V will always result in very dangerous current levels. The most severe result of an electrical shock is death.

The worst possible frequency for humans is 60 Hz, which is commonly used in utility power systems. Humans are about five times more sensitive to 60-Hz alternating current than to direct current. At 60 Hz, humans are more than six times as sensitive to alternating current than at 5000 Hz, and the sensitivity appears to decrease still further as the frequency increases. Above 100 to 200 kHz, sensations change from tingling to warmth, although serious burns can occur from higher RF energy.

At much higher frequencies (e.g., above 1 MHz), the body again becomes sensitive to the effects of an alternating electric current, and contact with a conductor is no longer necessary; energy is transferred to the body by means of electromagnetic radiation.

13.10.5.2 Body Impedance

Three components constitute body impedance: internal body resistance and the two skin resistances at the contact points with two surfaces of different voltage potential. One-hand (or single-point) body contact with electrical circuits or equipment prevents a person from completing a circuit between two surfaces of different voltage potential. Table 13.2 lists skin-contact resistances encountered under various conditions. It also shows the work-area surfaces and wearing apparel effects on the total resistance from the electrical power source to ground. This table can be used to determine how electrical hazards could affect a worker in varying situations.

Delayed reactions and even death can be caused by serious burns or other complications. The most dangerous current flow via the chest cavity is through the heart when the shock occurs in the time relative to the normal heart rhythm. This current may cause ventricular fibrillation, which is defined as repeated, rapid, uncoordinated contractions of the heart ventricles. Ventricular fibrillation that alters the heart's normal rhythmic pumping action can be initiated by a current flow of 75 mA or greater for 5 sec or more through the chest cavity.

TABLE 13.2
Human Resistance (Q) for Various Skin-Contact Conditions

Body Contact Condition	Dry (ohms)	Wet (ohms)
Finger touch	40,000–1,000,000	4,000–15,000
Hand holding wire	15,000–50,000	3,000–5,000
Finger–thumb grasp	10,000–30,000	2,000–5,000
Hand holding pliers	5,000–10,000	1,000–3,000
Palm touch	3,000–8000	1,000–2,000
Hand around 1.5-inch pipe or drill handle	1,000–3,000	500–1,500
Two hands around 1.5-inch pipe	500–1,500	250–750
Hand immersed	-	200–500
Foot immersed	-	100–300

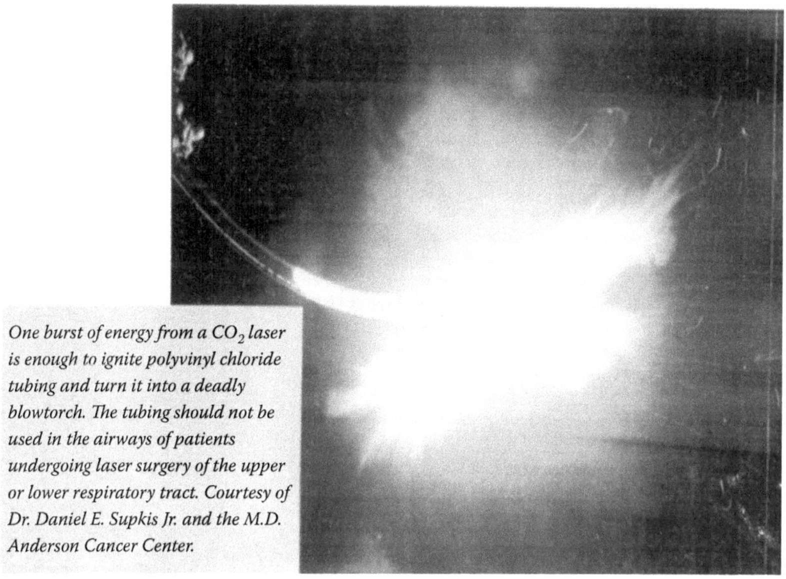

FIGURE 13.1 **(See color insert following page 164.)** Endotrachel tube on fire.

Bibliography

Dalziel, C.F., The effects of electric shock on man, *Industrial Radio Engineers Transactions on Medical Electronics*, May 1956.

Lee, R.H., Human electrical sheet, while an IEEE Fellow at E. I. duPont de Nemours & Co.

Lee, R.H., Electrical safety in industrial plants, *IEEE Spectrum*, June 1971. Anderson Cancer Center.

14 Practical Control Measures

Say, that is a good idea I can use.

This chapter for some will be the most useful; it is chiefly images of control measure solutions for different laser safety problems. Some are very simple and familiar. Of others, the reader may say, "Why didn't I think of that?" In addition to these examples is a listing of Web resources the reader may find useful to investigate:

14.1 WEB RESOURCES

www.hpa.org.uk/radiation/laser/
http://www.fda.gov/cdrh/
http://www.conformity.com/0210safety.pdf
www.ehrs.upenn.edu/programs/laser/laser_manual.html
www.osha.gov/SLTC/laserhazards
http://www.derm.ubc.ca/laser/eyesafety.html
http://chppm-www.apgea.army.mil/laser/Publications/Main.html#TG
web.princeton.edu/sites/ehs/laserguide/sec5.htm
www.ehs.uiuc.edu/rss/laser/tutorial.htm
www.rli.com/resources/stnd-lst.asp
www.prolaser.co.uk
www.ilda.wa.org/Laserist/Safety_2.html
repairfaq.ece.drexel.edu/sam/
radsafe.berkeley.edu/lasersafety.html
www.llnl.gov/es_and_h/esh-manual.html

American National Standard for the Safe Use of Lasers ANSI Z136.1 (2000), Laser Institute of America, ANSI, New York, TA1677.ASI2.

Henderson, A.R., *A Guide to Laser Safety* (1997), Chapman & Hall, New York, TA1677.H46.

Mathews, L. and Garcia, G., *Laser and Eye Safety* (1995), IEEE Press, New York, RE831.M38.

Winburn, D.C., *Practical Laser Safety* (1985), Marcel Dekker Inc., New York, TA1677.W560.

For useful resources on the Web, try:

Sam's Laser FAQ: Sam's FAQ is a personal Web site with lots of practical info. http://www.misty.com/people/don/lasersam.htm

LaserNet: LaserNet is the homepage for Rockwell Industries, makers of laser safety equipment. http://www.rli.com/

OSHA Laser Hazards: This government site has lots of useful technical information. http://www.osha-slc.gov/SLTC/laserhazards/index.html

This chapter is composed of examples of control measures — all applicable to the laser research environment.

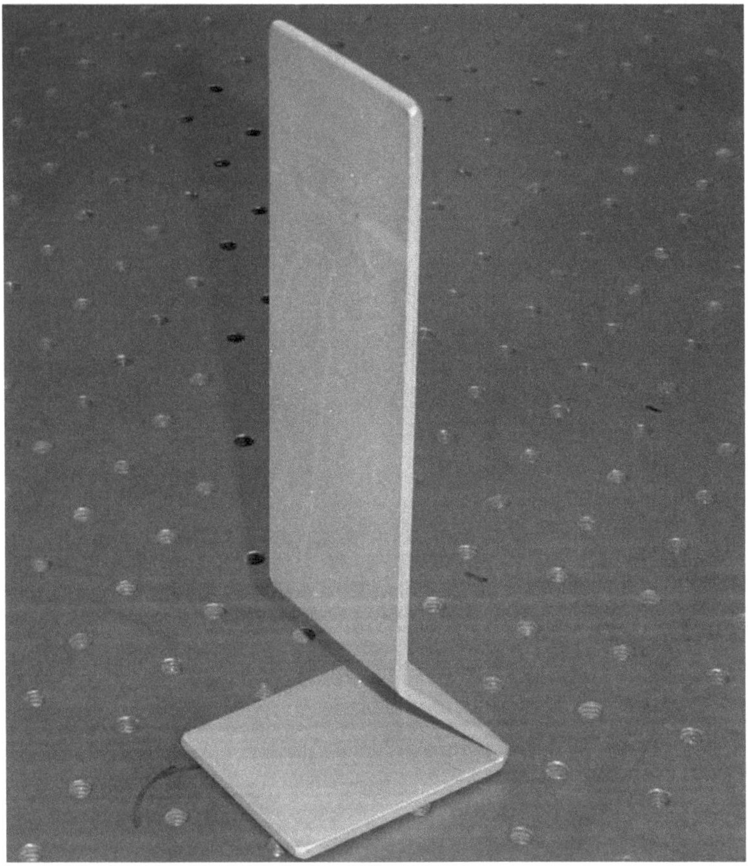

FIGURE 14.1 **(See color insert following page 164.)** Homemade beam block.

FIGURE 14.2 Magnetic strips applied to base.

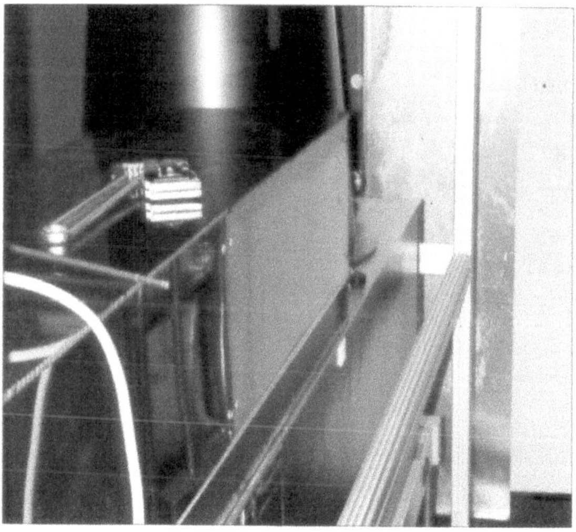

FIGURE 14.3 80/20 frame around optical table.

FIGURE 14.4 **(See color insert following page 164.)** View from inside enclosed table.

FIGURE 14.5 Table enclosure with view.

FIGURE 14.6 Table enclosure.

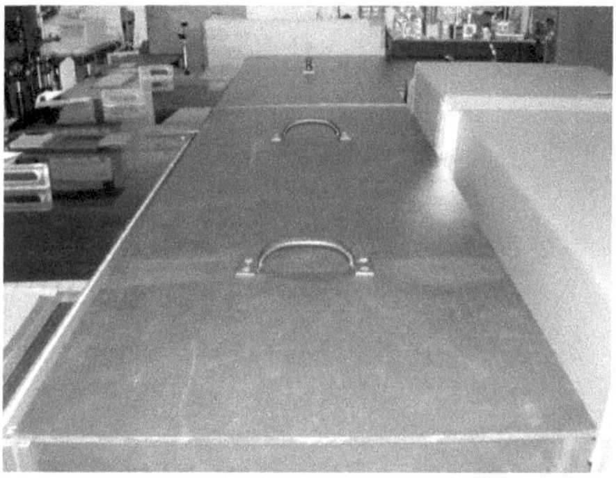

FIGURE 14.7 Enclosure lids.

FIGURE 14.8 Cables out of enclosure.

FIGURE 14.9 Multiperimeter guards.

FIGURE 14.10 Table block.

FIGURE 14.11 Clips holding perimeter guard (metal sheets).

FIGURE 14.12 Curtain with binder notebook.

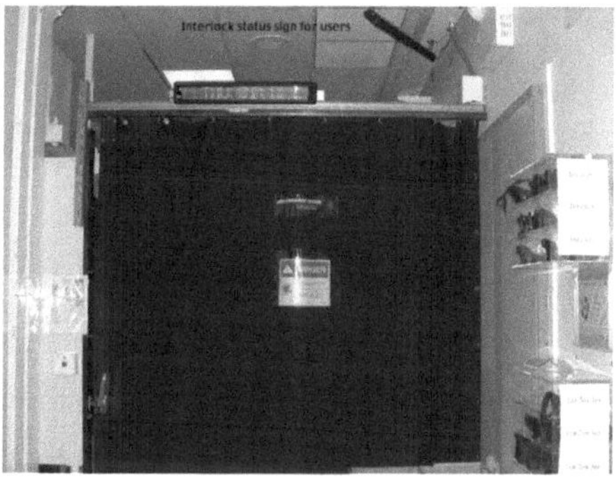

FIGURE 14.13 Curtain with scroll digital sing, eyewear holder.

FIGURE 14.14 Eyewear holder.

FIGURE 14.15 **(See color insert following page 164.)** Similar eyewear holder.

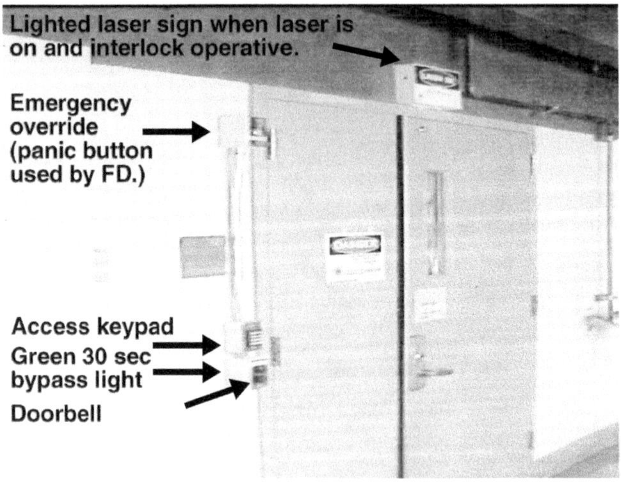

FIGURE 14.16 **(See color insert following page 164.)** Sign too high; binder on door.

FIGURE 14.17 Indicator light too high.

FIGURE 14.18 Labeled beam blocks.

FIGURE 14.19 **(See color insert following page 164.)** Acrylic enclosure.

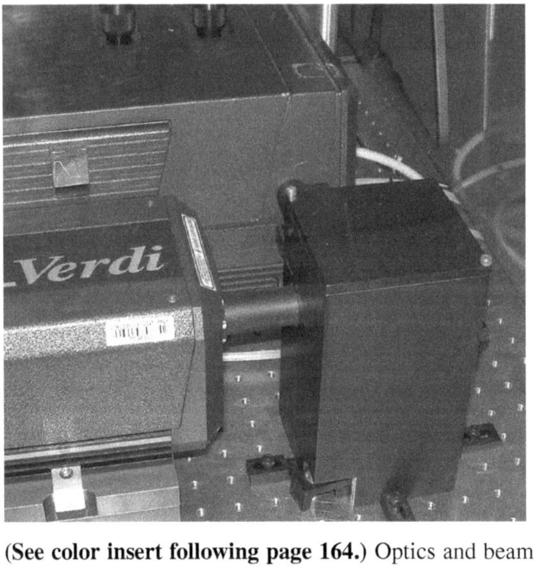

FIGURE 14.20 **(See color insert following page 164.)** Optics and beam path enclosed.

FIGURE 14.21 PVC beam tube.

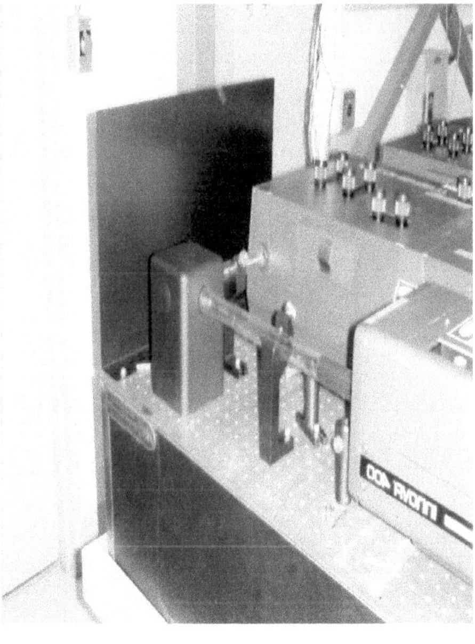

FIGURE 14.22 Beam tube and perimeter guard near door.

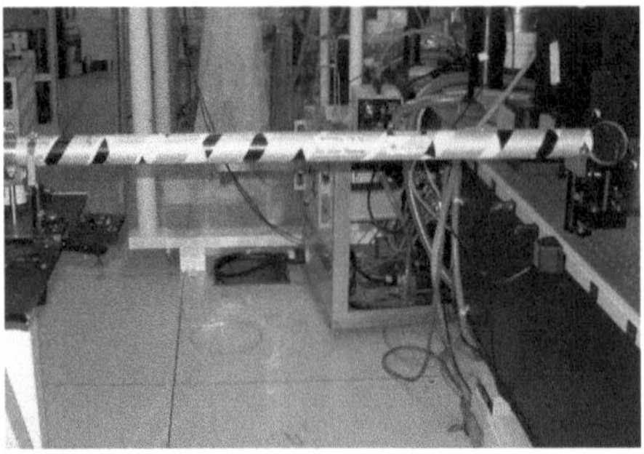

FIGURE 14.23 Tube across walkway.

FIGURE 14.24 Walkway swing arm in up position.

FIGURE 14.25 Walkway blocked.

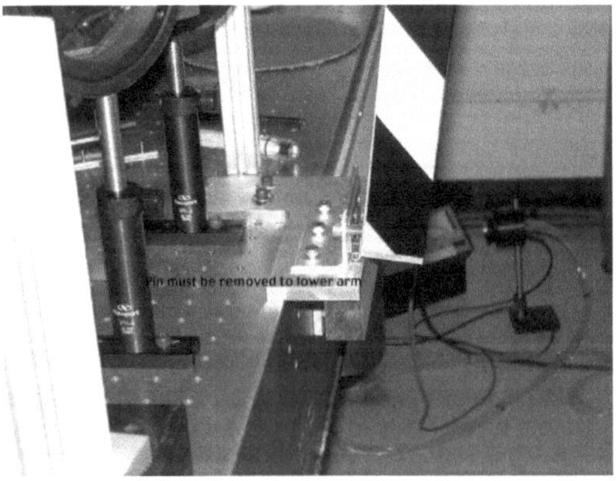

FIGURE 14.26 Pin holding swing arm up.

FIGURE 14.27 Warning lights and too many signs.

FIGURE 14.28 Crash button sign.

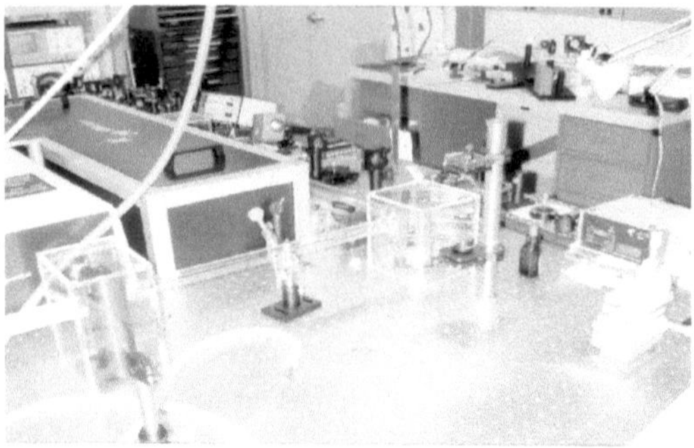

FIGURE 14.29 Tube keeping dust and fingers off.

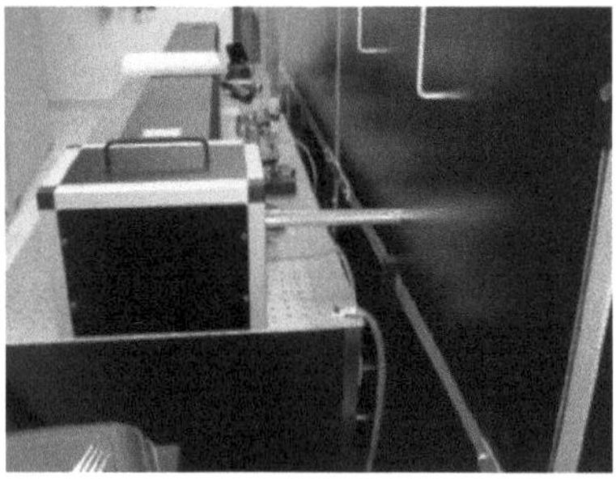

FIGURE 14.30 Enclosure and beam tube.

FIGURE 14.31 (**See color insert following page 164.**) Dye stain.

FIGURE 14.32 **(See color insert following page 164.)** Secondary container for dye pumps.

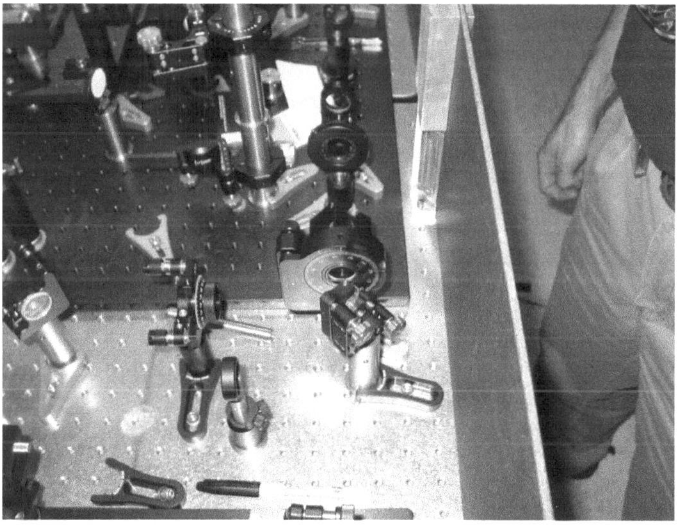

FIGURE 14.33 Perimeter guard.

FIGURE 14.34 Iris open, alignment aid.

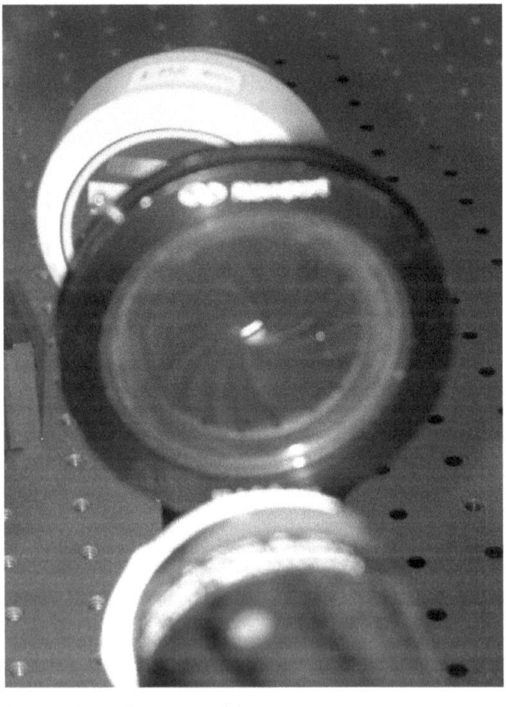

FIGURE 14.35 Iris smaller, alignment aid.

FIGURE 14.36 Is this your lab?

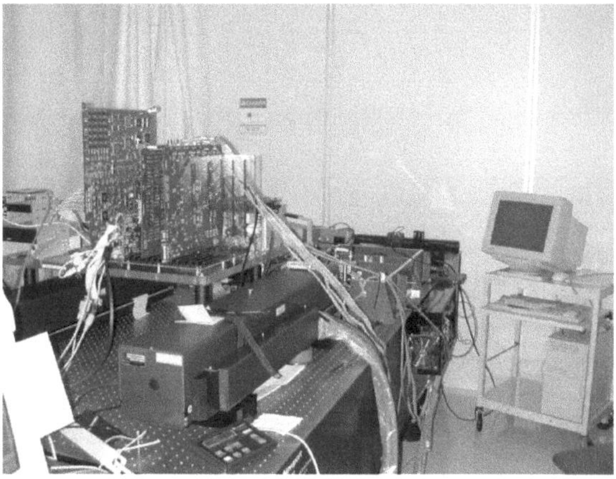

FIGURE 14.37 Another lab problem.

FIGURE 14.38 Scroll display over door, binder near door.

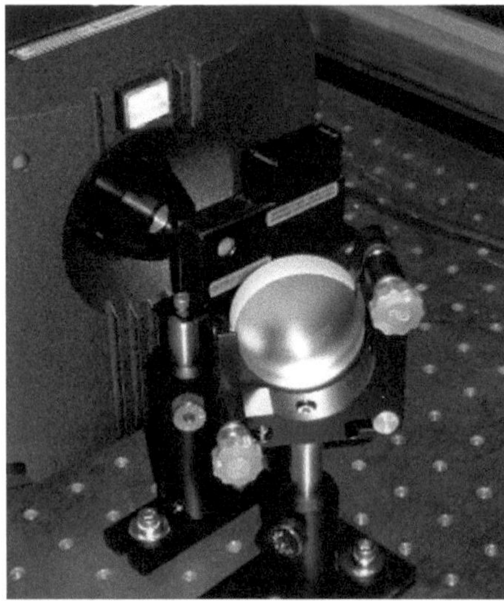

FIGURE 14.39 **(See color insert following page 164.)** External shutter.

FIGURE 14.40 Poor beam control.

FIGURE 14.41 Interlock system.

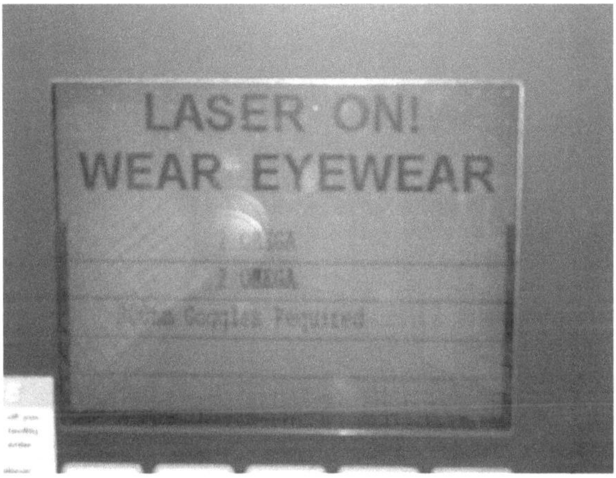

FIGURE 14.42 Laser status display PLC.

FIGURE 14.43 Laser diode in optical mount.

FIGURE 14.44 Polaroid paper.

FIGURE 14.45 Use of notice sign and flashing LED to indicate special administrative controls in use.

15 Laser Basics

15.1 INTRODUCTION

The laser, at least in concept, traces back to Albert Einstein in the early 1900s with his paper on stimulated emission. It took until 1960 for what we now call a laser to be demonstrated. A paper published by Townes and Schawlow in 1958 laid out the principles for the laser (light amplification of stimulated emission of radiation). In May 1960, American physicist Theodore Maiman built the first laser to successfully produce a pulse of coherent light, using synthetic ruby as the laser medium. The first continuously operating laser was achieved a few months later.

15.2 WHAT IS A LASER?

The word *laser* is an acronym for light amplification by stimulated emission of radiation. Figure 15.1 illustrates the total electromagnetic spectrum and wavelengths of the various regions.

The color, or wavelength, of light being emitted depends on the type of lasing material being used. For example, if a neodymium: yttrium aluminum garnet (Nd:YAG) crystal is used as the lasing material, light with a wavelength of 1064 nm will be emitted. Table 15.1 illustrates various types of material currently used for lasing and the corresponding wavelengths. Note that certain materials and gases are capable of emitting more than one wavelength. The wavelength of the light emitted in this case is dependent on the optical configuration of the laser. Lasers tend to be known by their medium.

Medium	Wavelength (microns)
Argon fluoride (excimer-UV)	0.193
Krypton chloride (excimer-UV)	0.222
Krypton fluoride (excimer-UV)	0.248
Xenon chloride (excimer-UV)	0.308 and 0.459
Xenon fluoride (excimer-UV)	0.351
Helium cadmium (UV)	0.325–0.442
Nitrogen (UV)	0.337
Helium cadmium (violet)	0.441
Krypton (blue)	0.476
Argon (blue)	0.488
Copper vapor (green)	0.510
Argon (green)	0.514
Krypton (green)	0.528

Frequency doubled Nd:YAG green	0.532
Helium neon (green)	0.543
Krypton (yellow)	0.568
Copper vapor (yellow)	0.570
Helium neon (yellow)	0.594
Helium neon (orange)	0.610
Gold vapor (red)	0.627
Helium neon (red)	0.633
Krypton (red)	0.647
Rohodamine 6G dye (tunable)	0.570–0.650
Ruby ($CrAlO_3$) (red)	0.694
Ti:sapphire (NIR)	0.690–0.960
Gallium arsenide (diode-NIR)	0.840
Nd:YAG (NIR)	1.064
Helium neon (NIR)	1.15
Erbium (NIR)	1.540
Helium neon (NIR)	3.39
Hydrogen fluoride (NIR)	2.70
Carbon dioxide (FIR)	9.6
Carbon dioxide (FIR)	10.6

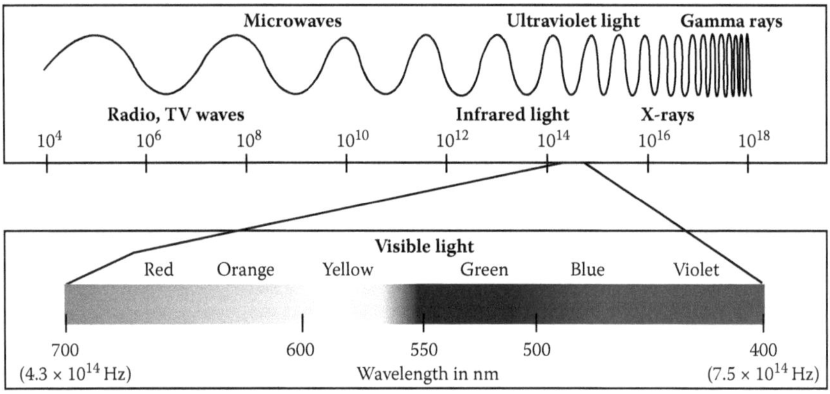

FIGURE 15.1 Electromagnetic spectrum.

15.3 LASER THEORY AND OPERATION

A laser generates a beam of very intense light. The major difference between laser light and light generated by white light sources (such as a light bulb) is that laser light is monochromatic, directional, and coherent. *Monochromatic* means that all of the light produced by the laser is of a single wavelength. White light is a combination of all visible wavelengths (400 to 700 nm). *Directional* means that the beam of light has very low "divergence" or spread. *Coherent* means that all the light is moving in the same direction and the waves are in "phase" with each other. A light bulb produces many wavelengths, making it incoherent (Figure 15.3).

TABLE 15.1
Common Lasers and Their Wavelengths

Laser Type	Wavelength (in nanometers)
Xenon chloride	308 and 459
Xenon fluoride	353 and 459
Helium cadmium	325–442
Copper vapor	511 and 578
Argon	457–528 (514.5 and 488 most used)
Frequency doubled Nd:YAG	532
Helium neon*	543, 594, 612, and 632.8
Krypton	337.5–799.3 (647.1–676.4 most used)
Laser diodes*	630–1550
Ti: sapphire	690–960
Nd: YAG*	1,064
Hydrogen fluoride	2,600–3,000
Erbium: glass*	1,540
Carbon monoxide	5,000–6,000
Carbon dioxide*	10,600

Light from conventional sources, such as a light bulb, diverges, spreading in all directions, as illustrated in Figure 15.2. The intensity may be large at the source, but it decreases rapidly as an observer moves away from the source.

In contrast, the output of a laser, as shown in Figure 15.4, has a very small divergence and can maintain high beam intensities over long ranges. Thus, relatively low-power lasers are able to project more energy at a single wavelength within a narrow beam than can be obtained from much more powerful conventional light sources.

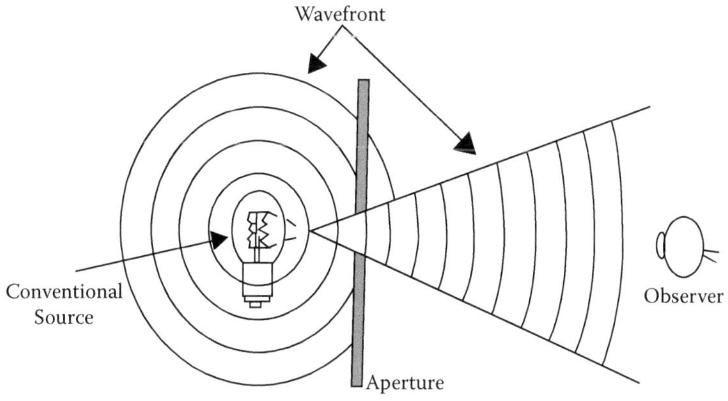

FIGURE 15.2 Divergence of conventional light source.

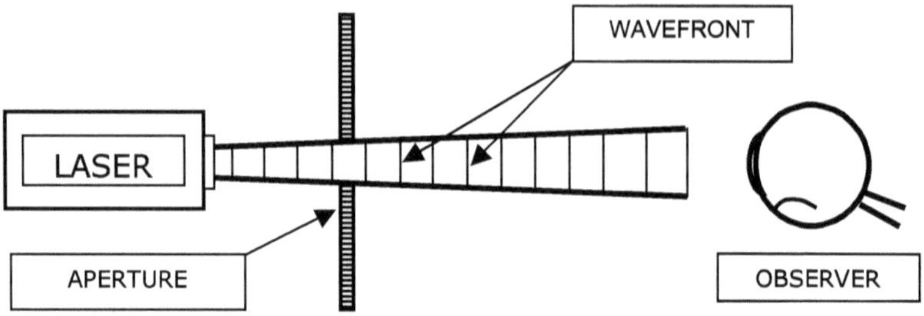

FIGURE 15.3 Divergence of laser source.

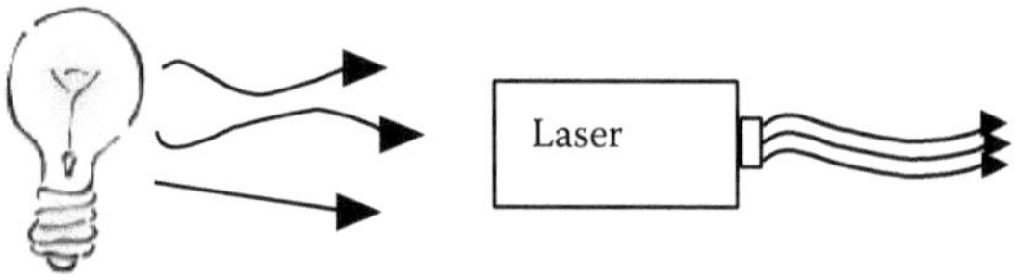

FIGURE 15.4 Incoherent light bulb versus coherent laser.

15.4 COMPONENTS OF A LASER

Figure 15.5 illustrates the basic components of the laser including the lasing material, pump source or excitation medium, optical cavity, and output coupler. The lasing material can be a solid, liquid, gas or semiconductor, and can emit light in all directions. The pump source is typically electricity from a power supply, lamp, or flashtube but may also be another laser.

The excitation source, sometimes referred to as the pump source, is used to excite the lasing material, causing it to emit light. The *optical cavity* contains mirrors at each end that reflect this light and cause it to bounce between the mirrors. As a result, the energy from the excitation medium is amplified in the form of light. Some of the light passes through the *output coupler*, usually a semi-transparent mirror at one end of the cavity.

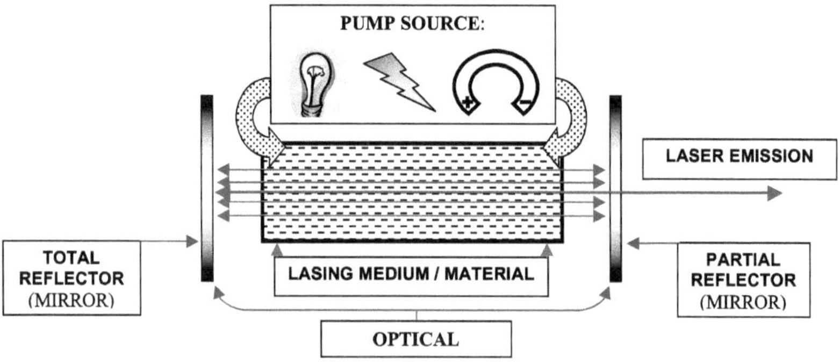

FIGURE 15.5 Basic laser diagram.

The excitation source, or pump source, provides energy to the medium, exciting the laser medium atoms such that electrons held within the atoms are elevated temporarily to higher energy states. The electrons held in this excited state cannot remain there indefinitely and drop down to a lower energy level. During this process, the electron loses the excess energy gained from the pump energy by emitting a photon. This is called spontaneous emission, and the photons produced by this method are the seeds for laser generation.

The photons emitted by spontaneous emission eventually strike other electrons in the higher energy states. *Eventually* is a very short time because of the speed of light and density of excited atoms. The incoming photon "knocks" the electron from the excited state to a lower energy level, creating another photon. These two photons are coherent, meaning they are in phase, of the same wavelength, and traveling in the same direction. This is called stimulated emission.

The photons are emitted in all directions, but some travel along the laser medium to strike the resonator mirrors and are reflected back through the medium. The resonator mirrors define the preferential amplification direction for stimulated emission. For the amplification to occur, there must be a greater percentage of atoms in the excited state than in the lower energy levels. This "population inversion" of more atoms in the excited state leads to the conditions required for laser generation.

The laser output may be steady, as in *continuous wave* (CW) lasers, or *pulsed*. A Q-switch in the optical path is a method of providing laser pulses of extremely short duration. The Q-switch may be a rotating prism, a pockels cell, a shutter device, or a radio-frequency (RF) gated quartz optical substrate used to create the pulse. Q-switched lasers may produce a high-peak-power laser pulse of a few nanoseconds' duration.

A CW laser has a steady power output, measured in watts. For pulsed lasers, the output generally refers to energy, rather than power. Radiant energy is a function of time and is measured in joules. Two terms are often used when measuring or calculating exposure to laser radiation. *Radiant exposure* is the radiant energy divided by the area of the surface the beam strikes. It is expressed in joules per square centimeter. *Irradiance* is the radiant power that strikes a surface divided by the area of the surface over which the radiant power is distributed. It is expressed in watts per square centimeter.

For repetitively pulsed lasers, the pulse repetition factor (PRF) and pulse width are important in evaluating biological effects.

Repetition Rate or Pulsing Frequency: The number of pulses per second from the laser, measured in Hertz. A laser producing 10 pulses per second is operating at a repetition rate of 10 Hz.

Pulse Energy: This value, measured in joules, is the total amount of energy in a laser pulse.

Pulse Duration or Pulse Width: The time duration of the pulse. Pulse duration is measured in nanoseconds but can be as fast as a femtosecond (10^{-15}) or attosecond (10^{-18})

Peak Power: A laser's peak power is its instantaneous output power during the laser pulse. Peak power is calculated in kilowatts as energy in joules divided by pulse duration in milliseconds. For example, a laser producing 5-J pulses of 2-msec pulse duration is producing 2.5 kW of peak power (5 J/2 msec = 2.5 kW). If the pulse duration is increased to 5 msec for the same 5-J pulse, the peak power will be reduced to 1 kW peak power.

Average Power: A pulsed laser's average power is measured in watts and is the average output power over a several seconds. The laser pulse energy multiplied by the laser's repetition rate determines the laser's average power output.

15.5 TYPES OF LASERS

The laser diode is a light-emitting diode that uses an optical cavity to amplify the light emitted from the energy band gap that exists in semiconductors (see Figure 15.6) It can be tuned to different wavelengths by varying the applied current, temperature, or magnetic field.

Gas lasers consist of a gas-filled tube placed in the laser cavity as shown in Figure 15.7. A voltage (the external pump source) is applied to the tube to excite the atoms in the gas to a population inversion. The light emitted from this type of laser is normally CW. If Brewster angle windows are attached to the gas discharge tube, some laser radiation might be reflected out the side of the laser cavity.

Dye lasers employ an active material in a liquid suspension. The dye cell contains the lasing medium. These lasers are popular because they may be tuned to several wavelengths by changing the chemical composition of the dye. Many of the commonly used dyes and liquid suspensions are toxic.

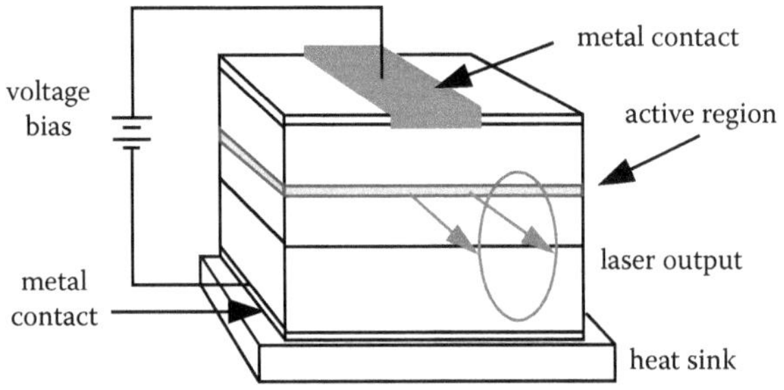

FIGURE 15.6 Semiconductor laser diagram.

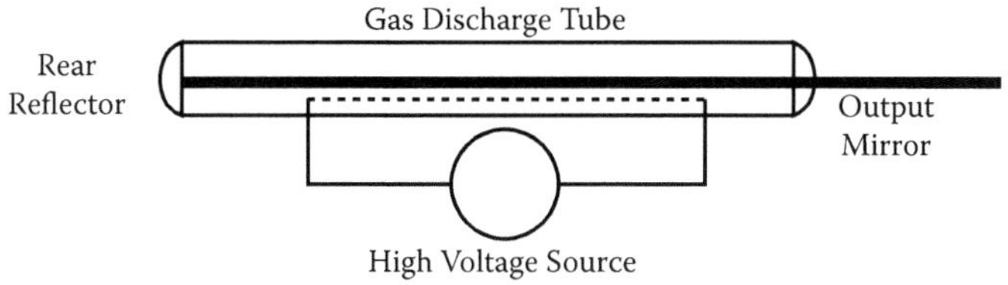

FIGURE 15.7 Gas laser diagram.

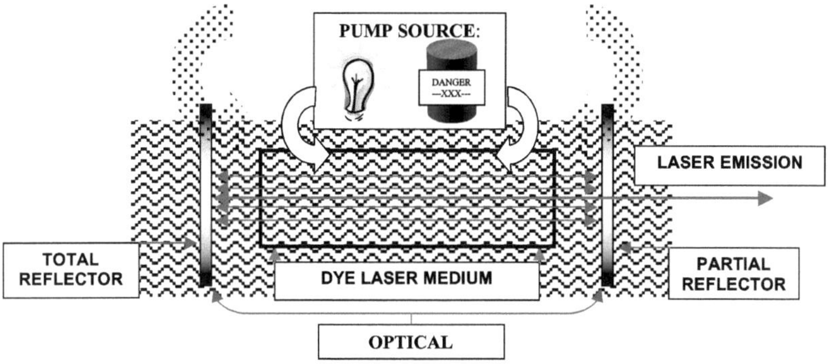

FIGURE 15.8 Block diagram of free dye laser.

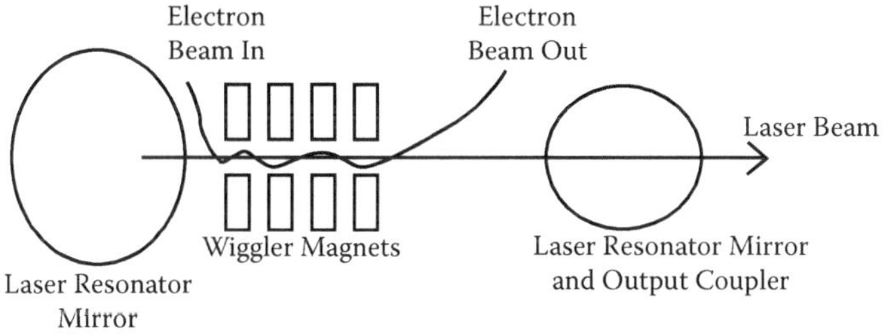

FIGURE 15.9 Block diagram of electron laser.

Free electron lasers (FELs) such as in Figure 15.9 generate wavelengths from the microwave to the x-ray region. They operate by having an electron beam in an optical cavity pass through a wiggler magnetic field. The change in direction exerted by the magnetic field on the electrons causes them to emit photons (from nLight Inc.).

Glossary

Absorption The reduction in amplitude for a beam of radiation incident in the medium through which it is propagated.

Access panel Part of the protective housing or enclosure that provides access to laser radiation when removed or displaced.

Accessible Emission Limit (AEL) The maximum accessible emission level permitted within a particular laser class.

Accessible optical radiation Optical radiation to which the human eye or skin may be exposed for the condition (operation, maintenance, or service) specified.

Active medium The material that acts as the light amplifier in a laser. It is typically excited by optical, chemical, or electrical means.

Administrative controls Safety measures of a nonengineering type such as key control safety training and warning notices.

AEL See *accessible emission limit.*

Alignment aids Alignment aids should be used whenever routine maintenance requires the alignment of beam path components.

Analogue meter Displays the power measured as a continuously varying function of the input to the detector.

Angular subtense (α) The visual angle subtended by the apparent source (including diffuse reflections) at the eye of the observer or at the point of measurement.

Aperture Any opening in a protective housing or other enclosure of a laser product through which laser radiation is emitted.

Aperture stop An opening that defines the area over which radiation is measured.

Apparent source The real or virtual object that forms the smallest possible retinal image. The concept of an apparent source is used in the extended wavelength region from 302.5 nm to 4000 nm since focusing by conventional lenses might be possible in this region.

Atom The smallest part of an element that can take part in a reaction. An atom consists of electrons orbiting a nucleus.

Audit A systematic critical examination carried out by qualified people to verify that appropriate standards are being set and met.

Automatic power reduction A feature of an optical fiber communication system by which the accessible power is reduced to a specified level within a specified time, whenever there is an event that could result in human exposure to radiation, for example a fiber cable break.

Beam attenuator A device to be introduced into the path of a beam that reduces the laser radiation to below a specified level.

Beam expander A combination of optical elements that increase the diameter of a laser beam.

Beam path component An optical component that lies on a defined beam path such as a beam steering mirror or a focusing lens.

Beam stop Any device that terminates a laser beam path.

By-products (degradation products) The material ejected during material processing as a result of radiation and material interaction causing breakdown products.

Cavity The optical assembly of a laser usually containing two or more highly reflecting mirrors that reflect radiation back into the active medium of the laser.

CEN Comite European de Normalisation (European Committee for Standardization).

Coherence Phenomenon in which light waves are "in-step." Correlation (constant phase angle) between the electric field of light at any point in space.

Collateral radiation Any electromagnetic radiation, within the wavelength range between 180 nm and 1 mm, except laser radiation, emitted by a laser product as a result of, or physically necessary for, the operation of a laser.

Collimated beam A "parallel" beam of light with a very small angular divergence or convergence. Beam cross-section remains quasi-constant with distance from the source.

Continuous wave (CW) An uninterrupted laser radiation output. The output of a laser that is operated in a continuous manner rather than pulsed mode, that is, a laser operating with a continuous output period equal to or greater than 0.25 sec.

Cryogenic Pertaining to very low temperatures.

Damage threshold The minimum power or energy density, which if incident on a surface will result in damage.

Defined beam path An intended path of a laser beam within a laser product.

Delivery system The optical system that takes laser radiation from the laser output and delivers it to the point where it can interact, for example, optical fibers and flight tubes.

Diffraction A phenomenon that causes interference fringes to appear when a beam of light passes through an aperture. The fringes are a result of the wave nature of light.

Diffuse reflection Destroys the collimated nature of the beam and occurs when the beam is incident on a nonmirror like surface that is not totally absorbent.

Digital meter Displays the power measured as numbers composed of digits.

Divergent beam A beam whose cross-section increases with distance from the source.

Electromagnetic radiation Energy produced by vibrating electric and magnetic fields. Electromagnetic radiation can be thought of as waves or streams of photons.

Electromagnetic spectrum The range of frequencies over which electromagnetic radiations are propagated. The spectrum ranges from short wavelengths such as gamma rays and X-rays, through visible radiation to longer wavelength radiations of microwaves, television, and radio waves.

Embedded laser product A product, which because of engineering features, limits the accessible emissions and has been assigned a class number lower than the inherent capability of the laser incorporated.

Emission duration The temporal duration of a pulse, of a train, or series of pulses or of continuous operation, during which human access to laser radiation could occur as a result of operation, maintenance, or servicing of a laser product.

Emission warning device Any warning device that gives an audible or visible warning that a laser system is switched on or capacitor banks of a pulsed laser are charged or have not been positively discharged.

Enclosed system System in which, during normal operation, the optical radiation is totally enclosed, for example, by light-proof cabinets, components, total internal reflection, or optical fiber cables and connectors.

End-user The person or organization using the optical fiber communication system in the manner the system was designed to be used. The user cannot necessarily control the power generated and transmitted within the system.

Engineering control Safety measures of a deliberate engineering design that should be used as the fundamental method of reducing exposure to radiation. A physical means of preventing access to radiation.

Excimer An *ex*cited di*mer*. A system of two similar atoms that, in their excited states, join together and form a molecule in an excited state.

Excited state An atom or molecule when it is in an energy level with more energy than in its normal, or "ground," state.

Exposure time The duration of a pulse, or series, or train of pulses or of continuous emission of laser radiation incident upon the human body.

Extended source A source is considered an extended source when the apparent source subtends a linear angle greater than α_{min} (unlike most laser sources). Extended source viewing conditions occur when laser radiation is reflected from a diffusing surface and the image formed by the reflected radiation on the retina of the eye exposed (apparent source) is larger than a certain minimum value.

Fail Safe A component whereby its failure does not increase the hazard; that is, it fails in a safe condition. In the failure mode the system is rendered inoperative or nonhazardous.

Frequency The number of cycles per unit time of an oscillation. Symbol: f Unit: Hz.

Fumes The airborne emission of the material breakdown during radiation or material processing or interaction.

Giant pulsed Pulsed lasers with pulse durations of $<10^{-9}$ sec.

Hazard Something with the potential to cause harm. The hazard can be to people, property, or the environment.

Hazard level The potential hazard at any accessible location within an optical fiber communication system. It is based on the level of optical radiation that could become accessible in reasonably foreseeable circumstances, for example, a fiber cable break. It is closely related to the laser classification procedure in BS EN 60825-1.

Human access The ability for any part of the human body to be exposed to hazardous laser radiation as emitted by an aperture or access gained by probe (given limits and definitions). Within a protective housing, human access can often be gained from reflecting surfaces inside the product or through openings in the protective housing.

Interlock A mechanical or electrical device to prevent the hazardous operation of a system.

Intrabeam viewing All viewing (and measurement) conditions in which the eye is exposed to laser radiation, other than extended source viewing. Examples are viewing of collimated beams and point-type sources.

IR Infrared region of the spectrum composed of wavelengths in the range of 700 to 10^6 nm.

Ionising radiation Radiation of sufficient energy to cause ionization (produce ions) of substances through which it passes, for example, gamma radiation and x-rays.

Irradiance The power of a laser averaged over the area of the beam (W m^{-2}).

Key control Class 3B and class 4 laser products should be key operated such that they can be protected against unauthorized use. A key can be other control devices such as magnetic cards.

Laser An acronym that describes a device that produces monochromatic light of high coherence: light amplification by stimulated emission of radiation.

Laser-associated hazard A hazard arising directly from the operation or installation of the laser or laser system alone.

Laser controlled area An area within which people are subject to control and supervision for the purpose of protection from laser radiation hazards.

Laser product Any product or assembly of components that constitutes, incorporates, or is intended to incorporate a laser or laser system and that is not sold to another manufacturer for use as a component (or replacement for such component) of an electronic product.

Laser safety officer A person who is knowledgeable in the evaluation and control of laser hazards and has responsibility for the oversight of such.

Laser system A laser in combination with an appropriate laser energy source with or without additional incorporated components.

Light (visible radiation) A form of electromagnetic radiation able to be detected by the human eye. Its wavelength range is between approximately 400 and 700 nm (far red to far violet).

Light Emitting Diode (LED) Any semiconductor pn junction device that is designed to produce electromagnetic radiation by radiative recombination in a semiconductor in the wavelength region from 180 to 1 mm. Radiation is produced primarily by the process of spontaneous emission, although some stimulated emission may be present.

Likelihood The probability or chance that a hazard will occur.

Limiting aperture The circular area over which irradiance and radiant exposure are averaged.

Luminous transmittance/Visible light transmittance (VLT) The amount of white light transmitted by the filter. Expressed as a percentage.

Maintenance The performance of those adjustments or procedures specified in user information provided by the manufacturer with the laser product that are to be performed by the user for the purpose of assuring the intended performance of the product. It does not include operation or service.

Manufacturer An organization or individual who assembles optical devices and other components in order to construct or modify an optical fiber communication system.

Maximum angular subtense (α_{max}) The value of angular subtense of the apparent source above which the MPEs and AELs are independent of the source size. α_{min} = 0.1 rad = 100 mrad.

Maximum Permissible Exposure (MPE) The level of radiation to which, under normal circumstances, persons may be exposed without suffering adverse effects. MPE levels are the maximum levels to which eye and skin can be exposed without injury and are related to wavelength, exposure duration, tissue type, and viewing conditions.

Meter A unit of length in the international system of units; currently defined as the length of a path traversed in vacuum by light during a period of 1/299792458 seconds. Typically, the meter is subdivided into the following units: centimeter (cm) = 10^{-2} m; millimeter (mm) = 10^{-3} m; micrometer (μm) = 10^{-6} m; nanometer (nm) = 10^{-9} m.

Microsecond 10^{-6} second (1 μs, one-millionth of a second).

Millisecond 10^{-3} second (1 ms, one-thousandth of a second).

Minimum angular subtense (α_{min}) The value of the angular subtense of the apparent source above which a source is considered an extended source. MPEs and AELs are independent of the source size for angular subtenses less than α_{min} = 1.5 mrad.

Mode locking The production of a very short pulse of laser radiation emitted from the laser as a result of one round trip of the laser cavity. Pulses are produced at regular intervals.

Monochromatic Light that is composed entirely of one color or frequency.

Nanosecond 10^{-9} second (1 ns, one thousand millionth of a second).

Nominal ocular hazard distance (NOHD) The distance at which the beam irradiance equals the appropriate MPE value. An indication of a "safe viewing" distance for certain criteria. An equivalent term for skin exposure is *hazard distance*.

Nonionizing radiation Radiation that does not posses sufficient energy to cause ionization in biological materials.

Operation The performance of the laser product over the full range of its intended functions. It does not include maintenance or service.

Optical Density (OD) A measure of a filter's ability to absorb laser radiation. It is dependent on laser wavelength, transmittance of the filter, and filter thickness; numerical value.

Optical fiber communication system An engineered assembly for the generation, transference, and reception of optical radiation arising from lasers in which the transference is by means of optical fiber for communication and control purposes.

Optical gain The optical amplification of radiation caused by the focusing properties of the eye or other photodetector.

Optical radiation Part of the electromagnetic spectrum composed of infrared, visible, and ultraviolet radiations.

Outcome The result of a hazard being realized. Also be known as *consequence* or *harm*. Can include financial loss, damage, injury, or even death.

Period The time for one complete oscillation, wave motion, or other regularly repeated process. Symbol: T It is the reciprocal of frequency.

Personal Protective Equipment (PPE) Safety equipment designed to protect a person from laser radiation and other hazards such as gloves, goggles, and ear defenders. Necessary when other methods (engineering controls, administrative controls) are not adequate.

Photometry Measurement of properties of visible radiation, for example, intensity. Applies only to a region capable of invoking vision in the human eye (~400 to 700 nm).

Photon A "particle" of electromagnetic radiation. It is believed that radiations can show wave behavior in some situations and particle behaviors in others.

Pockel's cell An electro-optical device used to rotate the plane of polarization of light by applying an electric field or pulse. Used inside a laser cavity to Q-switch or mode lock to produce a pulsed output.

Polarization The direction of vibration of the electromagnetic wave in the laser beam. Polarization may be plane, circular, elliptical, or random.

Population inversion The condition needed for lasing action whereby the number of atoms in an excited state is greater than the number of atoms in a lower energy state. Usually achieved by rapidly supplying energy to the active medium.

Process-associated hazard Hazard arising as a consequence of the particular use or application of a laser or laser product.

Protective enclosure A physical means for preventing human exposure to laser radiation unless such access is necessary for the intended functions of the installation.

Protective housing Those portions of a laser product (including a product incorporating an embedded laser) that are designed to prevent human access to laser radiation in excess of the prescribed AEL (generally installed by a manufacturer).

Pulse duration The time increment measured between the half peak power points and the leading and trailing edges of a pulse.

Pulsed laser A laser that delivers its energy in the form of a single pulse or a train of pulses. Under BS EN 60825-1, the duration of a pulse is less than 0.25 sec.

Q-switched laser A laser that emits short (~10 to 250 ns), high-power pulses by means of a Q-switch.

Q-switch The laser process in which lasing action is prevented while the population inversion increases. The increased energy level is then rapidly released to produce an enhanced energy in a very short pulse. A device for producing very short (~10 to 250 ns), intense laser pulses by enhancing the storage and dumping of electronic energy in and out of the lasing medium, respectively.

Quantum detector (photodetector) Detects radiation incident upon it via an element sensitive to the number of photons incident upon it. Examples of photodetectors are photodiodes and photomultiplier tubes.

Radian (rad) A unit of angular measure equal to the angle subtended at the center of a circle by an arc whose length is equal to the radius of the circle. 1 radian ~57.3 degrees; 2π radians = 360 degrees.

Radiance Radiant flux or power output per unit solid angle per unit area expressed in watts percentimeter squared per steradian ($W\ cm^{-2}\ sr^{-1}$).

Radiant energy Energy emitted, transferred, or received in the form of radiation. Unit: joules (J).

Radiant exposure Surface density of the radiant energy received, expressed in units of joules per centimeter squared (Jcm^{-2}).

Radiant flux Power emitted, transferred, or received in the form of radiation. Unit: watts (W). Also called *radiant power.*

Radiant intensity Quotient of the radiant flux leaving a source and propagated into an element of solid angle containing the direction, by the element of solid angle. Radiant intensity is expressed in units of watts per steradian (Wsr^{-1}).

Radiant power Power emitted, transferred, or received in the form of radiation, expressed in watts (W). Also called *radiant flux.*

Radiometry A branch of science that deals with the measurement of radiation. For the purposes of this standard, radiometry will be limited to the measurement of infrared, visible, and ultraviolet radiation.

Rayleigh scattering Scattering of radiation in the course of its passage through a medium containing particles whose sizes are small compared with the wavelength of the radiation.

Radiant exposure Total energy of radiation incident on a surface per unit area ($J\ m^{-2}$).

Radiation The emission of energy from a source, either as waves (light, sound) or as moving particles (beta rays, alpha rays).

Radiometry Measurement of properties of optical radiation (UV, visible, IR). Concerned with the power content of radiation, for example radiant power.

Reflectance The ratio of total reflected radiant power to total incident power. Also called *reflectivity*.

Refractive index (of a medium) Denoted by n, the ratio of the velocity of light in vacuum to the phase velocity in the medium. Also called *index of refraction*.

Repetitive pulse laser A laser with multiple pulses of radiant energy occurring in a sequence.

Retina The sensory membrane that receives the incident image formed by the cornea and lens of the human eye. The retina lines the inside of the eye.

Retinal hazard region Optical radiation with wavelengths between 0.4 and 1.4 μm, where the principal hazard is usually to the retina.

Reasonably foreseeable event The occurrence of an event that under given circumstances can be predicted fairly accurately and the occurrence probability or frequency of which is not low or very low. Examples of reasonably foreseeable events include fiber cable break, optical connector disconnection, operator error, and inattention to safe working practices.

Reflection The process in which radiation meeting a boundary between two media bounces back to stay in the first medium. Deviation of radiation following incidence on a surface.

Refraction The bending of light as it passes through a medium.

Remote interlock connector A connector that permits the connection of external controls placed apart from other components of the laser product.

Risk factor The product of the likelihood of a hazard occurring and the outcome or harm that arises as a result.

Safety interlock An automatic device associated with the protective housing of a laser product to prevent human access to class 3B or class 4 laser radiation when that portion of the housing is removed.

Scanning laser radiation Laser radiation having a time-varying direction, origin, or pattern of propagation with respect to a stationary frame of reference.

Scattering The spreading out of a beam of radiation as it passes through matter, reducing the energy moving in the original direction.

Scattered laser radiation Laser radiation that deviates from the beam path. Examples include unwanted secondary reflections from beam path components, deviant radiation from misaligned or damaged components, and reflections from a work piece.

Service The performance of those procedures or adjustments described in the manufacturer's service instructions that may affect any aspect of the product's performance. It does not include maintenance or operation.

Service panel An access panel that is designed to be removed or displaced for service.

Specular reflection A reflection that maintains the collimated and point-source characteristics of the source and occurs when the beam is incident on a mirror-like surface.

Secured enclosure An enclosure to which casual access is impeded by an appropriate means, for example, a door secured by a magnetically or electrically operated lock or latch, or by fasteners that need a tool to remove.

Shall The word *shall* is to be understood as mandatory.

Should The word *should* is to be understood as advisory.

Small source In this book, a source with an angular subtense at the cornea equal to or less than alpha-min (α_{min}), that is, ≤ 1.5 mrad. This includes all sources formerly referred to as "point sources" and meeting small-source viewing (formerly called point source or intrabeam viewing) conditions.

Small-source viewing The viewing condition whereby the angular subtense of the source, α_{min}, is equal to or less than the limiting angular subtense, α_{min}.

Solid angle The three-dimensional angular spread at the vertex of a cone measured by the area intercepted by the cone on a unit sphere whose center is the vertex of the cone. Solid angle is expressed in steradians (sr).

Source A laser or a laser-illuminated reflecting surface.

Spectator An individual who wishes to observe or watch a laser or laser system in operation and who may lack the appropriate laser safety training.

Specular reflection A mirror-like reflection.

Steradian (sr) The unit of measure for a solid angle. There are 4p steradians around any point in space.

Standard operating procedure (SOP) Formal written description of the safety and administrative procedures to be followed in performing a specific task.

Stimulated emission When a population inversion exists in a material, an emitted photon may cause the release of an identical photon from another atom or molecule in an excited energy state. Emission is thus described as "stimulated."

T_1 The exposure duration (time) at which MPEs based upon thermal injury are replaced by MPEs based upon photochemical injury to the retina.

T_2 The exposure duration (time) beyond which extended-source MPEs based upon thermal injury are expressed as a constant irradiance.

Threshold Limit (TL) In this standard, the term is applied to laser protective eyewear filters, protective windows, and barriers. The TL is an expression of the "resistance factor" for beam penetration of a laser protective device. This is generally related by the threshold limit (TL) of the protective device (expressed in Wcm^{-2} or Jcm^{-2}). It is the maximum average irradiance (or radiant exposure) at a given beam diameter for which a laser protective device (e.g., filter, window, barrier, etc.) provides adequate beam resistance. Thus, laser exposures delivered on the protective device at or below the TL limit beam penetration to levels at or below the applicable MPE.

Telescopic viewing Viewing an object from a long distance to increase it's visual size. These systems generally collect light through a large aperture, magnifying hazards from large-beam, collimated lasers.

Thermal effect An effect brought about by the temperature elevation of a substance (e.g., biological tissue). Photocoagulation of proteins resulting in a thermal burn is an example. The threshold radiant exposure is dependent upon the duration of exposure and heat transfer from the heated area.

T_{max} See *limiting exposure duration.*

t_{min} For a pulsed laser, the maximum duration for which the MPE is the same as the MPE for a 1-ns exposure. For thermal biological effects, this corresponds to the "thermal confinement duration" during which heat flow does not significantly change the absorbed energy content of the thermal relaxation volume of the irradiated tissue. (Example: t_{min} is 18 ms in the spectral region 0.4 to 1.05 mm and is 50 ms between 1.050 and 1.400 mm).

Tonometry Measurement of the pressure (tension) of the eyeball.

Thermal detector Detects radiation incident upon it via a temperature-sensitive element. Examples of thermal detectors are calorimeters and bolometers.

Tool A screwdriver, a coin, or another object that may be used to operate a screw or similar fixing means.

Transducer A device that converts a nonelectrical parameter, for example, light, into electrical signals, the variations in the electrical signals being a function of the input parameter.

Transmission The passage of radiation through a medium. If not all radiation is absorbed, that which passes through is said to be transmitted. Dependent upon wavelength, polarization, radiation intensity and transmitting material.

Transmittance The transmittance of a filter is the amount of light that it allows to pass. Expressed as a percentage.

Uncontrolled area An area where the occupancy and activity of those within is not subject to control and supervision for the purpose of protection from radiation hazards.

UV Region of the spectrum composed of wavelengths in the range of 100 to 400 nm. It can be further subdivided into three regions: UV-A (315 to 400 nm), UV-B (280 to 315 nm), and UV-C (100 to 280 nm).

Viewing window Visually transparent parts of enclosures that contain laser processes. It may be possible to observe the laser processes through the viewing windows.

Viewing portal An opening in a system that allows the user to observe the chamber. All viewing portals and display screens included as an integral part of a laser system must incorporate a suitable means to maintain the laser radiation at the viewing position at or below the applicable MPE for all conditions of operation and maintenance. It is essential that the material used for viewing portals and display screens be combustable or release toxic vapors following exposure to laser radiation.

Visible radiation Region of the spectrum extending from approximately 400 to 700 nm, that is, any optical radiation capable of directly causing a visual sensation in the human eye.

watt (W) The unit of power or radiant flux. 1 watt = 1 joule per second.

Wavelength The distance between two successive points on a periodic wave that have the same phase. Symbol:λ

Work practices Procedures used to accomplish a task.

X-rays Streams of X radiation. X radiation is an energetic form of electromagnetic radiation of wavelength range of 10^{-11} to 10^{-8} m. X radiation is ionizing radiation.

ACRONYMS

ACGIH American Conference of Governmental Industrial Hygienists

ANSI American National Standards Institute

AOM Acousto Optic Modulator

BSI British Standards Institution

CAA Civil Aviation Authority

CDRH Center for Devices and Radiological Health

CEN European Committee for Standardization

CENELEC European Committee for Electrotechnical Standardization

CIE International Commission on Illumination

CW Continuous Wave

DOH Department of Health

DTI Department for Trade and Industry

ICAO International Civil Aviation Organization

ICNIRP International Commission on Non-Ionizing Radiation Protection

IEC International Electrotechnical Commission

ILDA International Laser Display Association

IPEM Institute of Physics and Engineering in Medicine

ISO International Organization for Standardization

LASER Light Amplification by the Stimulated Emission of Radiation

LED Light Emitting Diode
LIA Laser Institute of America
LPA Laser Protection Adviser
LSMP Laser Safety Management Program
LSO Laser Safety Officer
MPE Maximum Permissible Exposure
NOHD Nominal Ocular Hazard Distance
NRPB National Radiological Protection Board
PCAOM Polychromatic Acousto Optic Modulator
WHO World Health Organization

Index

D

E

F

G

H

I

M

N

O

P

T

U